COORDINATION SCHEMES FOR DISTRIBUTED BOUNDARY COVERAGE WITH A SWARM OF MINIATURE ROBOTS: SYNTHESIS, ANALYSIS AND EXPERIMENTAL VALIDATION

THÈSE N° 3919 (2007)

PRÉSENTÉE LE 12 OCTOBRE 2007
À LA FACULTÉ INFORMATIQUE ET COMMUNICATIONS
GROUPE MARTINOLI
PROGRAMME DOCTORAL EN INFORMATIQUE, COMMUNICATIONS ET INFORMATION

ÉCOLE POLYTECHNIQUE FÉDÉRALE DE LAUSANNE

POUR L'OBTENTION DU GRADE DE DOCTEUR ÈS SCIENCES

PAR

Nicolaus CORRELL

Dipl. Elektroingenieur ETH
et de nationalité allemande

acceptée sur proposition du jury:

Prof. E. Telatar, président du jury
Prof. A. Martinoli, directeur de thèse
Prof. G. Kaminka, rapporteur
Prof. V. Kumar, rapporteur
Prof. J.-Y. Le Boudec, rapporteur

EPFL

ÉCOLE POLYTECHNIQUE
FÉDÉRALE DE LAUSANNE

Suisse
2007

Abstract

We provide a comparison of a series of original coordination mechanisms for the distributed boundary coverage problem with a swarm of miniature robots. Our analysis is based on real robot experimentation and models at different levels of abstraction. Distributed boundary coverage is an instance of the distributed coverage problem and has applications such as inspection of structures, de-mining, cleaning, and painting. Coverage is a particularly good example for the benefits of a multi-robot approach due to the potential for parallel task execution and additional robustness out of redundancy. The constraints imposed by a potential application, the autonomous inspection of a jet turbine engine, were our motivation for the algorithms considered in this thesis. Thus, there is particular emphasis on how algorithms perform under the influence of sensor and actuator noise, limited computational and communication capabilities, as well as on the policies about how to cope with such problems.

The algorithms developed in this dissertation can be classified into reactive and deliberative algorithms, as well as non-collaborative and collaborative algorithms. The performance of these algorithms ranges from very low to very high, corresponding to highly redundant coverage to near-optimal partitioning of the environments, respectively. At the same time, requirements and assumptions on the robotic platform and the environment (from no communication to global communication, and from no localization to global localization) are incrementally raised. All the algorithms are robust to sensor and actuator noise and gracefully decay to the performance of a randomized algorithm as a function of an increased noise level and/or additional hardware constraints.

Although the deliberative algorithms are fully deterministic, the actual performance is probabilistic due to inevitable sensor and actuator noise. For this reason, probabilistic models are used for predicting time to complete coverage and take into account sensor and actuator noise calibrated by using real hardware. For reactive systems with limited memory, the performance is captured using a compact representation based on rate equations that track the expected number of robots in a certain state. As the number of states explode for the deliberative algorithms that require a substantial use of memory, this approach becomes less tractable with the amount of deliberation performed, and we use Discrete Event System (DES) simulation in these cases.

Our contribution to the domain of multi-robot systems is three-fold. First, we provide a methodology for system identification and optimal control of a robot swarm using probabilistic models. Second, we develop a series of algorithms for distributed coverage by a team of miniature robots that gracefully decay from a near-optimal performance to the performance of a randomized approach under the influence of sensor and actuator noise. Third, we design an implement a miniature inspection platform based on the miniature robot Alice with ZigBee ready communication capabilities and color vision on a foot-print smaller than 2 x 2 x 3 cm^3.

Keywords: Swarm Robotics, Distributed Coverage, Multi-Robot Systems

Résumé

Cette dissertation étudie et compare des mécanismes de coordination pour le problème de couverture distribuée de contours (distributed boundary coverage problem) avec un essaim de robots miniatures. La comparaison se fonde sur l'expérimentation et la modélisation de robots à différents niveaux d'abstraction. La couverture distribuée de contours est un exemple du problème de couverture distribuée et elle trouve des applications dans l'inspection de structures, la détection de mines, le nettoyage ou encore la peinture. Les problèmes de couverture tirent un grand bénéfice des systèmes multi-robots grâce à leur parallèlisme et leur redondance qui leur donne une robustesse additionnelle. Les contraintes imposées par une application potentielle, l'inspection autonome d'une turbine de jet, servent de motivation pour les algorithmes présentés dans cette thèse. Ainsi, un accent tout particulier a été mis sur la performance des algorithmes sous l'influence du bruit des capteurs et des actuateurs, et de ressources de calcul et de mémoire limitées, mais également sur les diverses solutions permettant d'améliorer les performances dans de telles conditions.

La description formelle de la dynamique d'un système multi-robots est nécessaire pour la prévision et l'optimisation de l'exécution. Les algorithmes sont classifiés en diverses catégories: réactifs et délibératifs, ainsi que collaboratifs et non-collaboratifs. La performance de ces algorithmes va d'une couverture fortement redondante á une division proche de l'optimum du domaine, avec une croissance incrémentale des hypothèses et des exigences sur la plateforme robotique et l'environnement (d'aucune communication à une communication globale, et d'aucune localisation à une localisation globale). Tous les algorithmes sont robustes au bruit des senseurs et des actuateurs et se réduisent élégamment à l'exécution d'un algorithme randomisé lorsque le niveau de bruit et le nombre de contraintes matérielles augmente.

Bien que les algorithmes délibératifs soient entièrement déterministes, leur exécution réelle devient probabiliste en raison du bruit inévitable des senseurs et des actuateurs. Pour cette raison, des modèles probabilistes sont employés pour prévoir le temps de couverture tout en tenant compte du bruit des senseurs et des actuateurs calibré à l'aide de vrais robots. Pour les systèmes réactifs dont la mémoire est limitée, la performance est évaluée en utilisant une représentation compacte basée sur des équations différentielles qui modélise le nombre moyen de robots dans un certain état. Cette approche devient moins appropriée lorsque la quantité de mémoire utilisé pour une délibération devient substantielle. Dans de tels cas, on doit recourir à la simulation de systèmes d'événements discrets (DES).

La contribution de cette dissertation au domaine des systèmes multi-robots porte ainsi sur trois plans. Premièrement, elle fournit une méthodologie pour l'identification de systèmes et le contrôle optimal d'un essaim de robots en utilisant des modèles probabilistes. Deuxièmement, elle développe une série d'algorithmes pour la couverture distribuée par une équipe de robots miniatures qui se réduisent avec élégance d'une performance proche de l'optimum à celle d'une approche randomisée sous l'influence du bruit des senseurs et des actuateurs ainsi que des contraintes matérielles. Troisièmement, une plateforme miniature d'inspection basée sur le robot miniature Alice avec des possibilités de communication via ZigBee et de vision couleur, le tout integré dans un volume de moins de 2cm x 2cm x 3cm est conçue et mise en application.

Mot clés: essaim de robots, couverture distribuée, systèmes multi-robots

The modeling of multi-agent systems is of considerable interest for understanding chemical or biological processes, ranging from molecular self-assembly to self-organization in social insects. Despite some purely academical endeavors, this research finds immediate application in the design of drugs, communication systems, and inspired various computational methods for solving real world problems. The modeling of miniature multi-robot systems, although cross-fertilizing other domains, has only found a few immediate applications. From this perspective, I am very happy to have had the opportunity to further develop the methodological aspects of modeling miniature multi-robot systems using case studies, which might indeed lead to real-world applications, such as distributed inspection of engineered structures (this thesis) or low-stress animal control by robots integrated into the animal society. Although everyday use of these technologies lies still in the future, these case studies emphasize a general tendency: Miniature agents show only limited capabilities that support rather a reactive, distributed coordination approach instead of deliberative, centralized control. Although further miniaturization will soon replace current miniature robots with much more capable and reliable successors, we will also see much smaller robots with capabilities comparable to those of the robots considered in this thesis. This in turn will open new terrain for autonomous robotic operation, e.g., in the human body or inside micro-machinery. An intriguing direction is given by the ability of robotic platforms to successfully integrate into natural societies and alter their behavior, as it was demonstrated within the European project LEURRE. Here, miniature robots impregnated with cockroach pheromones were accepted by a swarm of natural cockroaches as congeners and were used to induce non-natural behaviors in the swarm. It is imaginable that this very simple concept of communication — luring natural agents by camouflage and mimicking their communication channels — will be soon available on even smaller scales. Then, self-locomoting nano agents might play the role of the Pied Piper of Hamelin by removing unwanted guests (such as virusses or cholesterol) from our bodies.

Keeping this outlook in mind, this thesis is focused on a single case study: the autonomous inspection of jet turbines by a swarm of miniature robots. This case study was boot-strapped by initial funding from the NASA Glenn Research Center at the California Institute of Technology. The tough constraints of such a scenario (miniaturization and local communication) motivated us to explore what is feasible with miniature robotic platforms. As technology developed during the last four years, adding new capabilities such as radio communication and color vision became feasible within the size constraints of the turbine scenario and with that the opportunity to also consider deliberative approaches for inspection. This development lead us to encompass a large variety of coordination approaches from reactive to fully planned. By highlighting the general aspects of the experiments conducted and the modeling techniques being used, insights gained in this thesis are not limited to the inspection scenario, but might be applicable to nano robotic swarms on the one hand, and large scale (considering both number of agents and the size of the individual) multi-robot systems on the other hand.

Acknowledgements

First of all I would like to thank my advisor Prof. Alcherio Martinoli for sharing his passion and enthusiasm for understanding and building collective systems with me. His understanding of swarm robotics from a system perspective, encompassing real hardware, algorithms and the analysis and permanent exposition to all of these three thrusts in his group has provided me with an unique learning experience for which I am very grateful. I would also like to thank my thesis committee Prof. Gal Kaminka, Prof. Vijay Kumar and Prof. Jean-Yves Le Boudec for carefully reviewing this manuscript and valuable input.

I express my gratitude to Dr. Edmond Wong and Dr. Johnathan Litt from the NASA Glenn research center for envisioning swarm robotic jet turbine inspection, to Dr. Roland Moser and Fernando Silverio from ALSTOM Power Systems for insights on gas turbine inspection and non-destructive evaluation methods, and to Dr. Francesco Mondada for helpful discussions on robotic inspection systems.

I am also grateful to Sean Bronée, Daniel Calico-Cañals, Gregory Mermoud, Teodora Miteva and Anne-Elisabeth Tran Qui who worked with me on their student projects or internships; they directly or indirectly contributed to this thesis by helping me explore various directions. I would like to thank in particular Patrick Amstutz, who collaborated on market based algorithms for distributed coverage during his master's thesis, Jonas Fritschy who has implemented the first version of the radio module used in this thesis during his semester project, Amanda Prorok for extending the probabilistic modeling methodology used in this thesis in a semester and master's project, and Samuel

Rutishauser who collaborated on the deliberative coverage of unknown environments during his semester and master's projects.

I thank Xavier Raemy from whom I learned a lot about hardware and firmware design, and whose support with the design of the communication and camera module has been crucial for the experimental validation of my results. I'm grateful to Peter Brühlmeier and André Badertscher for sharing their years of experience with me and their great help on all matters with electronic circuits, and to André Guignard for access to his workshop and helpful advice on all sorts of mechanical problems. Also, I would like to thank Dr. Masoud Asadpour and Dr. Gilles Caprari for their help in getting me started with the Alice platform, and Dr. Yuri Lopez de Meneses for his support in bootstrapping the SwisTrack project.

Finally, I would like to thank all the people at the SWIS group for their support and friendship during and beyond the time of this thesis. In particular Yvan Bourquin and Dr. Olivier Michel for his great support concerning all matters of Webots, Erika Raetz for keeping everything going, Denis Rochat for his help with the Condor computational cluster, and Christopher Cianci, Gregory Mermoud, Dr. Julien Nembrini, Jim Pugh, Thomas Lochmatter and Pierre Roduit for being great colleagues!

Nikolaus Correll, Lausanne, Autumn 2007

CHAPTER 1

Introduction

Multi-robot systems can be a competitive alternative to a single robot solution, as they offer a higher level of robustness due to redundancy and the potential for individual simplicity. The possibility of conducting work in parallel potentially allows for faster task execution, e.g., in a coverage or an exploration task. This property is even more striking when size constraints on the robotic platform do not allow for task completion with a single robot in acceptable time. Besides physical constraints such as miniaturization and locomotion that are specific to the environment, such a scenario poses numerous design challenges such as limited inter-robot communication, limited computation, and a limited energy budget. Also, noise on crude sensors and actuators makes the design of deterministic control systems difficult because the performance of the system as a whole is essentially probabilistic. Thus, algorithmic design should aim at optimizing the overall performance of the swarm, rather than an optimal sequence of (inter-)actions. Such an approach is particularly reasonable when the number of robots is large and requires algorithms to be statistically predictable.

Coverage has a variety of industrial, humanitarian, and military applications such as inspection, cleaning, painting, plowing, mowing, and de-mining. Coverage algorithms can be employed on various platforms ranging from ground vehicles to underwater and unmanned aerial vehicles. Besides applications in coverage itself, coverage algorithms might also be required for search and exploration applications.

The benefits and challenges of employing a multi-robot solution are well-illustrated by

Figure 1.1: The compressor section of a jet turbine. The internal dimensions are within the same order of magnitude as those of the miniature robotic systems used in this dissertation.

the automatic inspection of (jet) turbines (Figure 1.1), which is a promising commercial application (Wong & Litt 2004) and imposes severe constraints on the individual robotic platform. Due to these constraints, which can be considered extreme conditions for multi-robot coverage, lessons learned on such a case study might well be applicable for applications that require larger robotic platforms.

In order to minimize failures, jet turbine engines have to be inspected at regular intervals for evidence of internal distress such as cracking or erosion. This is usually performed visually using borescopes, as well as using ultra-sound and eddy current sensors (Federal Aviation Administration 1998), a process which is time-consuming and cost-intensive, in particular if it involves dismantling the turbine. One possible solution for accelerating and automating the inspection process is to rely on a swarm of autonomous, miniature robots that could be released into the turbine while still attached to the wing (Litt, Wong, Krasowski & Greer 2003). With the immediate prospect of reducing the down-time during regular inspection intervals, the final goal of such an approach is a distributed embedded system that allows for a shift from a schedule-based maintenance procedure to a condition-based procedure based on smart sensors and actuators (Garg 2004). Here the deployment of mobile sensors, rather than the installation of permanent sensors (Hunter 2003), is a compromise between increased system cost and the benefits from an in-situ inspection.

The focus of this work is on the analysis and synthesis of algorithms that coordinate a robot swarm, rather than developing specific solutions for locomotion or inspection for an individual robot in a turbine environment (see for instance Tâche, Fischer, Siegwart, Moser & Mondada (2007) or Fischer, Tâche & Siegwart (2007), and Friedrich, Galbraith & Hayward (2006), respectively, and references therein). Nevertheless, experimentation with real hardware (Figure 1.2) is undertaken and it serves both as validation and motivation for algorithm development. Consequently, emphasis is on robustness with respect to sensor and actuator noise of the minimalist platforms in use. Overcoming the challenges imposed by the turbine inspection scenario that dramatically limits possible designs of robotic sensors, can pave the way for other similar applications in the inspection of engineered or natural structures such as tanks, pipes, networks of galleries or airplane surfaces.

In the remainder of this chapter, we summarize the design challenges that require the development of algorithms and models considered in this thesis. We then briefly describe the experimental setup used and the algorithms developed. The chapter is concluded with a summary of objectives and contributions of this thesis.

1.1 Design Challenges for Miniature Inspection Systems

The narrow environment of a turbine imposes a series of constraints that drastically influence the design choices for the robotic platform and potential coordination algorithms:

- Miniaturization can be considered as the toughest constraint. Miniaturization significantly limits the choice of potential actuators, sensors, and available energy. While the trend goes towards further miniaturization of sensors and actuators, it seems that available energy will become more and more a bottleneck on miniature embedded systems. This will in turn limit not only the overall movement autonomy but also on-board computational power and communication.

- Energy limitations might be overcome by providing the robots with tethers, which would be also useful for easily removing broken or stuck robots from the turbine. Tethers, however, have the disadvantage of requiring stronger actuators because the robot has not only to self-locomote but also to pull the — potentially entangled — tether that might quickly outweigh the robotic platform, in particular if it is to be robust enough for manual removal of the robots. In a distributed system, the entangling of tether cables is even more likely and imposes additional constraints on path-planning algorithms.

Figure 1.2: A simplified mock-up of a jet turbine being inspected by a swarm of minia-ture robots show-cased during the Swiss-wide Festival "Science-et-Cité" in Spring 2005. Photo ©Alain Herzog.

- Due to the shielded and narrow structure of the turbine, which might act as a Faraday cage, communication is limited to short range. For the same reason, closed-loop control of the system by an outside supervisor (agent) is essentially unfeasible.

- Reliable locomotion in a highly structured, 3-dimensional, upside-down environ-ment poses tremendous mechanical challenges.

The algorithms and analysis presented in this dissertation experimentally tackle miniaturization, energy limitations, and limited range communication, although no lo-comotion principles other than wheeled differential-drive robots are explored. Besides physical constraints, the inspection task also presents various algorithmic challenges (Correll, Cianci, Raemy & Martinoli 2006), which are not the subject of this disserta-tion:

- Potentially redundant sensory information provided by the robot swarm needs to be fused and correlated to the location within the turbine where it was recorded.

- The (three-dimensional) data recorded within the environment needs to be ana-lyzed, e.g., for detecting flaws (potentially using an expert system).

Figure 1.3: Coordination schemes developed in this thesis have drastically different assumptions on the individual robotic platform and the system as a whole. From left to right: the basic Alice robot, the basic Alice with a radio-module and additional computational power, the basic Alice with radio and camera module. Background: a Telos mote that can serve as a base station and repeater.

- Appropriate control commands need to be synthesized and send to the robot swarm in order to achieve a desired collective behavior: for instance, for inspecting more closely a certain region of the structure.

1.2 Experimental Setup

Our robotic inspection nodes (Chapter 3) are based on the Alice miniature robot (Caprari & Siegwart 2005), which we extended by a communication module and a camera module, each with a dedicated micro-controller (see Figure 1.3). The communication module allows robots to share coverage progress among each other and to transmit sensory data to a base station. The camera module allows for a visual inspection of the environment, and an identification of bar-codes in the environment, which is required by some of the algorithms considered in this dissertation.

The robots operate in a 2D mock-up of a turbine with the blades as vertical extrusions. This setup allows us to model a series of real-world constraints such as unreliable communication (due to the absorption of the signal by the metal structure of the environment), miniaturization constraints, and sensor and actuator noise.

1.3   Modeling Multi-Robot Systems

The various algorithms studied in this thesis have in common that the robot and environmental states can never be estimated with certainty. Thus, a deterministic algorithm
with provable performance guarantees is reduced to the performance of a probabilistic
algorithm. In chapter 4 we show how models capture the *probability* of being in a certain state of the system; this probability can be derived for reactive controllers and for
controllers with a limited amount of memory. We also show that for controllers with
memory, the state space becomes quickly untractable and is unfeasible to enumerate and
to analyze with closed-form equations. In this case simulating the system equations using
probabilities that are carefully calibrated from the real robotic system yields valuable
predictions.

1.4   Distributed Coordination Schemes for Multi-Robot Inspec-
tion

We consider three classes of coordination approaches that are drastically different in the
control paradigm used and in their requirements on the individual robotic platform.

In Chapter 5, we consider a purely reactive approach that has minimal requirements
on the robotic platform (low-bandwidth, local communication, no localization). It uses
self-organization as a coordination paradigm and simple reactive heuristics for coverage
(Correll & Martinoli 2004*a*). Local infrared-communication is used for increasing dispersion of the robots in the environments (Correll, Rutishauser & Martinoli 2006). In this
scenario, the camera can potentially be used for inspection, but off-line processing for
mapping sensory and image data to the location where they were recorded is required.

In Chapter 6 we consider deliberative approaches: In Section 6.2 robots create topological maps of the environment (Correll, Rutishauser & Martinoli 2006) but do not
collaborate or have a global coordinate frame. We extend this approach by implicit
collaboration in Section 6.3. The algorithm requires sufficient bandwidth for sharing
maps among the robots (Rutishauser, Correll & Martinoli 2007), which are then used
by individual planning.

Finally the algorithms presented in section 6.4 require the environment to be known
in advance, which allows for nearly-optimal partitioning the environment among the
robots in a distributed fashion. This partitioning is achieved by market-based algorithms
(Lagoudakis, Markakis, Kempe, Keskinocak, Koenig, Kleywegt, Tovey, Meyerson & Jain
2005) where robots "bid" on parts of the environment they are willing to cover. By
continuous re-auctioning, robots can make up for slower or failed robots.

In Chapter 7 the three approaches are compared with respect to their performance, hardware requirements and the modeling techniques used for assessing the system performance.

1.5 Objectives of this Dissertation

The performance of a multi-robot inspection system is a trade-off between available capabilities of the individual robotic node (e.g., communication, localization), the reliability of the individual robotic node (e.g., sensor and actuator noise), and the reliability of the inspection sensor (e.g., probability of false-positives). This dissertation aims to quantitatively and qualitatively address this trade-off with experimenting with real robots and modeling the system at various levels of abstraction.

By highlighting the general properties of the algorithms and models used, the lessons learned on a particular case study aim to contribute to a general methodology for synthesis and analysis of multi-robot systems, possibly consisting of a large number of units and characterized by severe limitations at the individual robot level.

1.6 Contribution of this Dissertation

Using a case study concerned with distributed boundary coverage by a swarm of miniature robots, we show that an optimal configuration of hardware and software of a multi-robot system is dependent on constraints such as the available size, energy, computation and available time. This optimal configuration can be evaluated by modeling the system at a higher abstraction level and carefully taking into account its probabilistic aspects (sensor and actuator noise, robot controller) and calibrating model parameters on experimental data from a (sub-)set of the robotic system.

Moreover, we provide a modeling framework for probabilistic modeling of reactive and deliberative coordination mechanisms. A suite of non-collaborative and collaborative algorithms for distributed coverage is used as example. Close agreement among different model abstraction levels is achieved by identifying probabilistic elements in the various algorithms considered in this thesis, by calibrating their parameters and importing them into the model.

Finally, the experimental work presented provides a first-of-its-kind implementation of a team of 40 miniature robots endowed with wireless radio communication, and a color camera on a foot-print smaller than 2cm x 2cm x 3cm.

Chapter Summary

- Multi-Robot Coverage has a variety of commercial and humanitarian applications such as the inspection of engineered structures, de-mining, and environmental monitoring. As coverage can be conducted in parallel, it is well suited for implementation on a multi-robot system.

- A series of algorithms with increasing requirements on the individual robotic platform is compared for distributed coverage of environments with arbitrary cellular decompositions. The presented algorithms have been developed for coping with real-world constraints such as sensor and actuator noise and limited communication. They are validated on teams of real miniature robots and using realistic simulation.

- This dissertation contributes to a general methodology for modeling self-organized robotic systems. Using numerous selected self-organized boundary coverage experiments, it is shown how to systematically identify key parameters governing the interactions of the swarm members among themselves and the environment, and how to derive an abstract and compact mathematical representation for macroscopic properties of a robot swarm.

- We have developed and implemented a team of 40 miniature robotic inspection robots, able to self-localize, to communicate via a wireless link, and record color images, with an overall footprint less than 2cm x 2cm x 3cm, as well as an experimental setup that carefully models aspects of the turbine scenario.

CHAPTER 2

Background

This chapter reviews the current state of the art in distributed coverage and modeling of large-scale distributed robotic systems, which are the main research thrusts addressed in this dissertation. The chapter concludes with a brief review of related work that situates the contribution of this thesis at a system level, i.e. a distributed autonomous inspection system, which involves research on human-swarm interaction and sensor-fusion, development of suitable sensors for inspection and locomotion schemes that are appropriate for navigation in target environments.

2.1 Distributed Dispersion, Exploration and Coverage

This section reviews work on distributed coverage and related disciplines, such as dispersion and exploration. Unlike dispersion (Section 2.1.1) and exploration (Section 2.1.2), which have their origins mainly in applications for autonomous robots, coverage (Section 2.1.3) path planning is a classic robotic application, which aims at calculating trajectories that lead to complete coverage of a manifold by an end-effector with specific kinematics, e.g. as required by a paint job in a factory. Immobile robots with rigid links used in manufacturing, however, provide a much higher reliability for predicting their end-effector's position than does an autonomous, mobile robot operating in a dynamic and potentially unknown environment. For these reasons, research that aims at applying path planning theory to autonomous robots has been mostly of theoretical nature and/or is based on simulation in order to fulfill the requirements on robot capabilities (perfect sensors and actuators, global localization), and few studies consider coverage using the swarm robotics paradigm. On the contrary, if high-level planning and reactive behavior are well separated, i.e. a robot can detect and autonomously recover from failure, area coverage can indeed be abstracted as some sort of graph exploration or coverage problem, which allows for leveraging algorithms from graph theory and for corresponding analytical insight.

2.1.1 Dispersion

A statically dispersed swarm can serve as a communication backbone (McLurkin & Smith 2006), distributed sensor (Schwager, McLurkin & Rus 2006), or guiding facility (Payton, Estkowski & Howard 2003, Kumar, Rus & Singh 2004) for higher level agents (robots or humans). Depending on the scenario, dispersion might thus be a prerequisite for coverage.

McLurkin & Smith (2006) report an experiment where 108 *Swarmbot*[1] robots are dispersed in an indoor environment of around $280m^2$ by solely relying on local range and bearing information (update rate 4Hz) and local communication. The fully scalable, distributed algorithm is able to maintain the connectivity of the swarm, and allows the deployed swarm to be used for navigation by using information about the network topology. In contrast to the experiment of McLurkin & Smith (2006), Howard, Parker & Sukhatme (2006) use approximately 80 robots in a mapping/dispersion experiment. A small team of larger robots (around 6) with extensive navigation capabilities was used to map an indoor environment using an off-line SLAM algorithm. The resulting map was then used by a centralized controller to determine optimal deployment positions for a swarm of smaller robots, which were guided in teams of three to six robots to their deployment positions. Local range and bearing was achieved using on-board vision and a bar-coded beacon mounted on the robots. The experiment shows that hierarchical control might be an effective solution for dealing with large number of robots.

From the perspective of provable properties of the dispersion algorithm, Chaimowicz, Michael & Kumar's (2005) contribution is noteworthy. Chaimowicz et al. present a distributed control scheme that is based on implicit functions for deploying a system in the environment. The controller is provably convergent, and has been validated by a team of four robots, albeit relying on global localization in the experiment. More recently, Schwager et al. (2006) presented experimental results with 50 *Swarmbots* that are dispersed in the environment by moving towards the centroid of an online Voronoi partitioning of the environment, which is constructed using local range and bearing

---

[1]http://www.irobot.com

between robots. Schwager et al. then show that the proposed distributed algorithm achieves a dispersion that is optimal with respect to sampling a specific probability density function (a light source and photo-sensors are used in the experiments for creating and sampling the probability density function, respectively). The completeness of this analysis is dependent upon the accuracy of robotic sensors and actuators, however, which may prevent convergence in some scenarios.

2.1.2 Exploration

Exploration is closely related to dispersion, as robots must distribute themselves to maximize the rate at which the environment is explored. However, unlike in dispersion, robots do not remain stationary at their deployment positions, and additional behavioral components such as collective movement might be required to maintain coordination. Exploration is typically used in unknown environments where robots are not pre-endowed with a map.

Zlot, Stentz, Dias & Thayer (2002) have experimentally studied a market-based coordination scheme for exploration using a team of ten robots exploring an indoor environment and dividing the task using an auction algorithm (see also Zlot & Stentz 2006). Although the system takes advantage of a centralized unit, it is not necessary for successful exploration as auctions can be held locally by each robot (at cost of potential redundancy). A key challenge in auction-based approaches is to find the right trade-off between solution quality and computational and communication burden: in order to find optimal allocations among a robot team, the robots would need to negotiate all possible permutations of task-robot allocations (Berhault, Huang, Keskinocak, Koenig, Elmaghraby, Griffin & Kleywegt 2003), requiring an exponential number of messages passed among the team. In practice however, near-optimal results can be achieved by sequentially auctioning tasks to the robots (Dias & Stentz 2000).

Burgard, Moors, Stachniss & Schneider (2005) presents a centralized architecture for explicit collaboration by trading off travel cost and information gain in order to distribute the robots in the environment. In particular, Burgard et al. (2005) study the influence of limited communication range. Finally, researchers from other research communities (e.g. Albers & Henzinger 2000) abstract the robotic exploration problem as exploration of unknown graphs, which allows for calculation of upper and lower bounds on the performance of a series of exploration strategies.

2.1.3 Coverage

Robotic coverage shares many aspects with robotic exploration; both involve robots dispersing to the peripheries of the environment. However, exploration usually requires only remote sensing of all the boundaries of the environment, while coverage may be required over the entire area. In coverage, robots are typically equipped (in theory) with some kind of end-effector (e.g., mowing or vacuum cleaning) or low-range inspection device (e.g., for land-mine detection) and perform some sort of cellular decomposition of the environment that is used to plan coverage trajectories, see (Choset 2001) for an overview. The cellular decomposition can either be established off-line, which requires knowledge of the environment, as in (Jäger & Nebel 2002), or on-line using on-board sensors. Butler, Rizzi & Hollis (2001) presents a coverage algorithm requiring only bumper sensors and is therefore limited to rectangular environments, whereas (Acar, Choset, Zhang & Schervish 2003) uses long range sensors. Also, literature distinguishes between approaches that plan the robot's trajectories off-line (Zheng, Jain, Koenig & Kempe 2005), and those that plan trajectories on-line, in which case the environment needs not to be known in advance (Butler et al. 2001, Acar et al. 2003). There exist hybrid approaches that require initial knowledge of the environment but perform dynamic (re-)planning (Williams & Burdick 2006, Jäger & Nebel 2002). Another important axis is also, whether robots will make up for potential failures of other robots as in e.g. (Hazon, Mieli & Kaminka 2006) or (Rekleitis, New & Choset 2005) where the environment does not need to be known in advance, or (Hazon & Kaminka 2005) in known environments.

In (Rekleitis et al. 2005), an auction-based algorithm is used for arbitrating coverage among the robots. Hazon et al. (2006) presents a multi-robot coverage algorithm that builds upon the provably complete and optimal STC algorithm for covering grid-like environments from Gabriely & Rimon (2001). In Hazon et al.'s (2006) work, each robot constructs parts of a minimal spanning tree of the environment and keep track of the status of robots that have crossed their path, assuming global communication and localization. Spanning trees are shared such that no redundancy occurs if none of the robots fail. This policy is non-optimal as robots might "cut-off" each other and lead to a non-uniform distribution of work (see also Hazon & Kaminka 2005). Hazon & Kaminka consider optimal spanning trees of known environments that are divided among the robots based upon their initial position on the spanning tree. Finally, Svennebring & Koenig (2004) presents an ant-inspired multi-robot coverage algorithm where robots leave traces in the environment, which can then be used by other robots for implicit collaboration.

Notice that all of the above work concerning multi-robot coverage, except (Jäger & Nebel 2002), has been conducted in simulation.

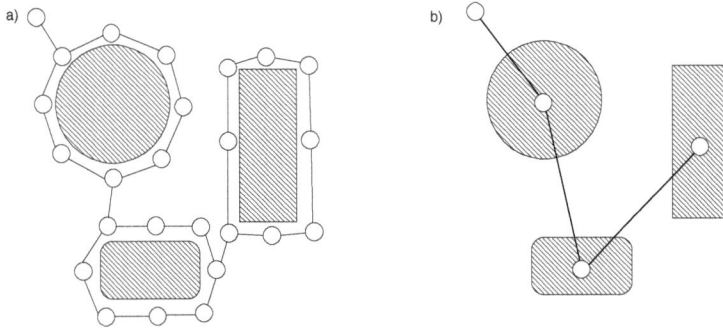

Figure 2.1: Complete coverage of the boundary of all objects requires traversing all edges of the graph to the left, or traversing all vertices of the graph to the right.

2.1.4 Multi-Robot Boundary Coverage

The boundary coverage problem is an instance of a graph coverage problem, where the graph is constructed so that its vertices cover only areas close to the boundary of objects in an enclosed space (Figure 2.1, *a*). In a dual representation (as in this dissertation), objects themselves can be considered as vertices of a graph and circumnavigation of the objects, i.e. boundary coverage, corresponds to covering a vertex (Figure 2.1, *b*). The distributed boundary coverage problem has applications in various potential robotic tasks, such as inspection or maintenance of structures, and has been first formulated by Easton & Burdick (2005). Easton & Burdick's (2005) work is mostly focussing on appropriate cellular decompositions for the boundary coverage problem, but also discusses near-optimal heuristics for distributing tasks among agents. In (Williams & Burdick 2006) these algorithms are extended by re-planning for increased robustness.

Chaimowicz et al. (2005) develops decentralized controllers for robot swarms to generate specific two-dimensional patterns defined by smooth functions generating a potential field. The controllers are provably stable for a class of boundaries (Hsieh & Kumar 2006). Hsieh, Loizou & Kumar (2007) then develop a methodology for generating potential functions that lead the robot swarm to *orbit*, i.e. circumnavigate the boundaries of the potential functions. As the work of Williams & Burdick, Hsieh & Kumar's algorithm require global localization for navigation.

2.2   Models for Swarm-Robotic Systems

For guiding the design process of self-organized robotic systems, a formal understanding of the relation between individual and collective behavior is desirable. Abstract models that capture this relation could help avoiding costly and time-consuming experiments, and might yield a priori insight into a specific system design.

2.2.1   Deterministic Models

A large amount of literature concerning large-scale, distributed, multi-robot systems originates from a system dynamics and control perspective. Results include algorithms and proofs of convergence of distributed controllers for flocking (Jadbabaie, Lin & Morse 2003), consensus (Olfati-Saber & Murray 2004, Ren, Beard & Atkins 2005), and optimal sensor distribution for sampling a given probability density function (Cortés, Martínez, Karatas & Bullo 2004).

The bases for these analyses are usually simplified motion models for the individual robots, artificial potential fields (Reif & Wang 1999), and graph structures that model neighborhood relations of the multi-robot system. Tight assumptions are made on the sensorial capabilities and the motion models of the robots in order to fit into the chosen mathematical framework (holonomic point robots with perfect sensors and actuators), but analysis methods have significantly improved in recent years, and the tendency is going from unrealistic synchronous swarms to asynchronous ones (Liu, Passino & Polycarpou 2003), from holonomic to non-holonmic point robots (Tanner, Jadbabaie & Pappas 2005), and towards accounting for sensor and actuator noise (Gazi & Fidan 2005).

Another perspective on modeling of multi-robot systems stems from classical robotic motion planning for addressing coverage tasks. Here, the environment is segmented using cellular decomposition (Choset 2001) and then partitioned among the robots. Computational geometry and graph theory provide the basis for algorithmic analysis (see for instance (Choset 2000) and (Rekleitis, Lee-Shue, New & Choset 2004) for the single and multi-robot cases, respectively).

Finally, operational research opens a perspective for addressing the multi-robot task allocation problem (which can be cast into a multi-robot coverage problem when considering partitions of the environment as tasks) and a valuable taxonomy and overview is provided by Gerkey & Matarić (2004). For distributed coverage Williams & Burdick (2006) and Zheng et al. (2005) both propose a centralized, near-optimal solution for partitioning an environment among a team of robots using constructive heuristics for the *k-rural postman problem* or the *n-binpacking problem*, respectively.

2.2.2   Probabilistic Models

Analysis using artificial potential fields or similar methods currently allows for analysis of only a subset of behavior-based control approaches in which the behavior of a robot does not change (i.e. the artificial potential field is the same for all robots). Models that keep track of the population dynamics of the swarm, i.e. the average number of robots in a certain state at some time, instead usually do not take into account the spatial distribution of the robots. Nevertheless, such models have shown strong quantitative agreement with a series of real-world robotic case studies where the performance metric is non-spatial (or can be formulated as such): object clustering (Martinoli, Ijspeert & Mondada 1999), collaborative manipulation (Martinoli, Easton & Agassounon 2004), and inspection (Correll, Rutishauser & Martinoli 2006). For foraging (Lerman, Jones, Galstyan & Matarić 2006), object aggregation (Agassounon, Martinoli & Easton 2004), and robot aggregation (Correll & Martinoli 2007a) good agreement has been obtained for model prediction and realistic simulation.

Probabilistic population dynamics models are derived by describing the individual robots behavior and environmental states with Markov chains (i.e. probabilistic finite state machines). State transition probabilities are calibrated with simple heuristics (see for instance Martinoli et al. 2004) or found using system identification based on experimental data (Section 4.4). The Markov chains can then be transcribed into a system of difference (Martinoli et al. 2004) or differential (Lerman, Martinoli & Galystan 2005) equations (one for each state) that summarize the average state transitions and thus track the average number of robots in each state. In many cases, interactions among the robots lead to state transition probabilities that are a function of the number of robots in other states, and thus yield a system of difference/differential equations that are non-linear.

Probabilistic modeling has also been used to address the robot task allocation problem by Agassounon et al. (2004) and Lerman et al. (2005). Here, the contribution from Agassounon et al. is of particular interest for swarm robotic coordination as task allocation in their work is based on threshold-based algorithms, which are inspired by division of labor in ant colonies (Bonabeau, Dorigo & Theraulaz 1999), and can be a competitive alternative to market-based solutions (Kalra & Martinoli 2006).

Although even simple systems become quickly analytically untractable, steady state analysis can be performed numerically, and stability and convergence can be determined using simple heuristics, e.g. phase diagrams (Strogatz 2000), or a reachability analysis of the state space can be performed using numerical computation toolboxes (Berman, Halasz, Kumar & Pratt 2006).

2.2.3   Hybrid Models

Hybrid systems are composed of both continuous and discrete components. The former are typically associated with physical first principles, whereas the latter are associated with logic switches, such as a Finite State Machine of a robot controller (see Balluchi, Benvenuti, Engell, Geyer, Johansson, Lamnabhi-Lagarrigue, Lygeros, Morari, Papafotiou, Sangiovanni-Vincentelli, Santucci & Stursberg's (2005) survey of the topic).

A robotic system becomes hybrid as soon as the behavior of an individual robot follows some rule based logic that switches between different behaviors whereas its physical state (e.g., position) is described by continuous values. Other instances of hybrid systems are the combination of continuous probability density functions (e.g., for modeling the position and discrete logic on individual level) or the combination of deterministic and probabilistic models (e.g., deterministic models for describing the kinematics of a system and probabilistic models for describing the behavioral state distribution of the system).

Indeed, deterministic models (Section 2.2.1) have classically modeled the spatial aspect, i.e. the distribution of the robots in the environment, whereas probabilistic models usually model the behavioral aspect of the system, i.e. the proportion of robots in a certain state. Both times, modeling needs to make assumptions such as either uniform distribution of robots and objects in the environment as commonly used by population dynamics models, or extremely simple behavior and perfect sensors/actuators for the individual robots as it is the case for deterministic models.

So far, only few hybrid modeling approaches for modeling multi-robot systems exist. Milutinovic & Lima (2006) show how an agent population can be controlled to move to a particular location using a centralized controller; this is accomplished by formulating the system as an optimal control problem on partial differential equations that describe the density function of the robots in the environment and the proportion of robots moving in one out of three different directions.

Berman et al. (2006) use a similar approach as Martinoli et al. (2004) and Lerman et al. (2005), and describe the population dynamics of a swarm of agents which is collectively looking for a resting site using differential equations (rate equations). The model is hybrid, as the swarm switches between different behavior sets with different dynamics at the macroscopic level and specific continuous control laws at the microscopic level.

Hybrid probabilistic models seem to be among the most promising for combining population dynamics with spatial dynamics of the swarm.

2.2.4 Multi-Level Modeling and Simulation Techniques

Given a distributed robotic system embedded in a dynamic environment, which has an almost infinite parameter space, ranging from the individual robot's controller and its morphology to features of the environment, key parameters need to be identified for describing a particular metric of interest with sufficient accuracy. Following the principle of parsimony (Occam's razor), the level of detail of a model can be gradually decreased. This yields drastically reduced experimental/simulation time and a more compact representation.

At the lowest abstraction level the system can be represented by realistic, embodied simulation, which faithfully reproduces body morphology, sensor features and placement, as well as physical constraints of the robots and the environment. Instances of such simulators are *Player/Stage/Gazebo* (Vaughan & Gerkey 2007) or *Webots* (Michel 2004), which has been used in this thesis.

Raising the level of abstraction, algorithmic properties of a system can be maintained by multi-agent simulation, where the deliberative parts of an individual robot controller are faithfully represented and have access to a common world model. Emphasis is on high-level interaction among the agents, rather than accurately modeling physics, kinematics, sensor and actuators. Instances of such simulators are *Swarm* (Minar, Burkhart, Langton & Askenazi 1996), *TeamBots* (Balch 1998) or *SPADES* (Riley & Riley 2003). At a higher level, some properties of a real system are intentionally replaced by average quantities. For instance, the robot speed together with its sensor range and the morphology of an object are abstracted by a constant probability for encountering this object at every time step which allows for simulating the system as a series of interacting, stochastic automata (Martinoli et al. 1999, Martinoli et al. 2004). Martinoli et al.'s (2004) implementation requires the metric of interest to be non-spatial, the distribution of the robots to be uniform in the environment, and synchronous simulation of the robot swarm. It reaches its limitations, however, when time scales in the system are very different (some events happen several orders of magnitude more often than others, e.g.) or when asynchronicity has a key impact on system performance. In these cases, the simulation can be more efficiently implemented as non-spatial discrete-event system (DES) simulation. The advantage of DES simulation is that the time of the next event in the system is computed beforehand and the simulator can skip simulation time in which no event happen. DES simulations have found wide-spread use for simulation of communication networks, and a large body of work on efficient implementation as well as open-source frameworks (see for instance the network simulator NS-2[2]) exist. In communication networks, however, computational units are classically non-mobile.

---

[2]http://www.isi.edu/nsnam/ns/

Available DES simulation frameworks are thus not directly applicable for implementing agent-based representations where agents follow the "perceive-plan-act" paradigm, see also Hybinette, Kraemer, Xiong, Matthews & Ahmed's (2006) discussion on the subject.

At the highest abstraction level, the macroscopic level, a robotic swarm can be described using difference equations (Martinoli et al. 2004) that keep track of the ratio of individuals in a certain state or the fraction of time the system as a whole spends in a certain state (Section 4.1). This approach is similar to population models commonly used in biology (Camazine, Deneubourg, Franks, Sneyd, Theraulaz & Bonabeau 2001), and has its origin in physical master equations that describe the probability of a system to occupy each one of a discrete set of states.

Martinoli et al. (2004) experimentally investigate the relation and potential mismatch between macroscopic analysis by difference equations and microscopic simulation for a swarm-robotic case study (the stick-pulling experiment). Microscopic simulation tends to yield better matching with macroscopic equations when the number of interactions in the system or the number of agents are large.

The approach of Martinoli et al., which explicitly simulates every individual agent and thus allows also studying heterogeneous swarms with no additional computational effort (Li, Martinoli & Abu-Mostafa 2004), reaches its limitations when the number of agents is large. A computational efficient, exact representation, although for differential equations, is the Gillespie algorithm (Gillespie 1977) which randomly generates possible trajectories for the underlying Master equations (see also Rathinam, Petzold, Cao & Gillespie (2003) and Gibson & Bruck (2000) for computational more efficient implementations of the original algorithm).

As the dynamics of the macroscopic model are exactly reproduced — but for low number of interactions or small number of agents — the Gillespie approach does not allow for lowering the abstraction level of the model in the microscopic representation or studying heterogenous teams, which is a strength of Martinoli et al.'s (2004) approach. Indeed, the microscopic level allows for studying additional details, e.g. communication loss, which are not captured by the macroscopic model, without requiring realistic simulation.

In practice results obtained from one or the other abstraction level / model should be treated with care, and the choice of a particular model has to be decided on a case-by-case basis and with respect to the aspect of the system one is interested in.

2.3   Robotic Inspection Systems

This section briefly reviews the system aspect of multi-robot systems for inspection or environmental monitoring and key technologies that might eventually enable the turbine inspection application, which is the motivating case study of this thesis.

The use of multiple robots for inspection and monitoring for the sake of safety, execution speed and robustness has been recurrently promoted in the literature. Petriu, Whalen, Abielmona & Stewart (2004) for instance present a series of heterogenous, networked robotic units for complex environmental monitoring tasks, and discusses issues in human-swarm interaction (see also Hinic, Petriu & Whalen (2007) for a study on human-computer symbiotic control of a robot swarm using a brain-computer interface), sensor fusion, and distributed coordination. Kumar et al. (2004) develops a scenario in which a team of autonomous agents penetrates a hazardous scenario and provide sensory information (such as temperature or air quality) as well as physical guidance through obstructed view (e.g., due to smoke) to a human team. Interestingly, navigation deadlocks are explicitly taken into account and immobile units are foreseen to serve as static sensor nodes and information relay.

2.3.1   Turbine Inspection

Using a team of robots specifically for the turbine inspection scenario was first articulated by Litt et al. (2003). Turbines currently used in large-scale commercial aircrafts have a length up to 7.5 meters (e.g., Pratt & Whitney PW4000, Rolls Royce Trent, General Electric GE90). The air-flow into the turbine is pre-compressed by a fan and directed into the compressor turbine. The compressed air is then enriched with fuel and burned in the combustion chamber, providing forward thrust. The resulting high-pressurized, hot gas is directed then through a further turbine, absorbing part of the thrust for driving the turbo fan, and the compressor turbine. Due to the extreme operation conditions imposed by a wide range of throttle requirements (take-off and landing vs. cruise) together with extreme variations in temperature and pressure in different operating altitudes and climatic regions, jet turbine engines are subject to extreme tear-and-wear ranging from burn-out of turbine blades due to hot gas from the combustion chamber (Garg 2004), and damage of the blade tips in compressor or turbine section due to rub with the outer shell, arising from thermal expansion of the materials that leads to reduction of the clearance between blade tips and turbine body, which is kept to a minimum for performance reasons (Melcher & Kypuros 2003), to foreign object damage (FOD) by objects from outside being sucked into the compressor or loose parts from within the turbine. There already exist different methods for assessing the engine health-state (Federal Aviation

Administration 1998), without actually taking measurements from within the turbine and hence increase on-wing time. For instance, analyzing the exhaust gas yields together with performance data such as the shaft-speed collected during flight, evidence of blade failures, compressor failure, foreign object ingestion damage, or seal erosion. Also, the analysis of oil condition by means of infra-red spectroscopy that can be performed within the engine itself, can indicate engine conditions by measurements of concentration of thermal and oxidative degradation products, water content, or fuel dilution. The above mentioned processes are associated with tremendous cost for the airline industry due to the down-time of equipment, and hence every possibility to enhance on-wing inspection leading to longer intervals for engine tear-down is highly sought after by the commercial airline industry (Garg 2004). Multi-robot systems are thus an interesting option for inspection of various sub-systems of the turbine, e.g. for inspection of blade mount-holes by navigating on the bore, inspection of tip clearance by navigating on the inside of the outer hull, or inspection of the combustion chamber.

2.3.2 Nondestructive Evaluation

A large body of work exists on Nondestructive Evaluation (NDE), often also referred to as Nondestructive Inspection (NDI) or Nondestructive Testing (NDT) using hand-held devices or tele-operated mobile robots. An interesting observation that is common to NDE sensors is the fact that sensors are not reliable and might produce false-positives and false-negatives test results, which advocates a multi-robot solution with a healthy amount of redundancy in coverage.

The most common sensors[3] for NDE base on the eddy current or on ultra-sound transducers, which require, however, application of contact gel as the difference in signal speed between air and matter make analysis of the obtained measurements difficult. Eddy current transducer are made of coils of around 3mm diameter, which allows to detect cracks down to a length of 0.1mm. For increasing spatial resolution of the sensor, coils can also be used in an array.

Siegel & Gunatilake (1997) developed a robotic platform for semi-autonomous inspection of aircraft skins using eddy-current sensors and visual inspection. Siegel & Gunatilake also discuss aspects of acceptance of new technology for inspection in safety critical domains. Sánchez, Vázquez & Paz (2005) present a robotic unit for semi-autonomous visual inspection of welding seams on the outer hull of large ships. Song, Wu & Kang (2004) presents a semi-autonomous inspection system for boiler tubes based on magnetic flux leakage and ultra-sound measurements, and Cruz & Ribeiro (2005) developed a robot for inspection of the bottom and walls of large oil tanks. Interestingly, NDE sen-

---

[3]according to a personal communication from Fernando Silverio, ALSTOM Power Service

sor technology allows for being used also on rather small miniature robots. Friedrich et al. (2006) presents a mobile robotic unit with magnetic wheels that allows for ultra-sound inspection of ferro-magnetic structures with a footprint of about 9cm x 9cm, whereas eddy-current sensors are available as hand-held units of the dimension of a pen (Siegel & Gunatilake 1997).

All of the above mentioned applications (including the turbine inspection scenario) have in common that the inspection unit usually needs to be multiple orders of magnitude smaller than the inspection domain, which further motivates the use of a multi-robot system.

Chapter Summary

- The field of multi-robot coverage has received considerable attention from the research community. However, only a few systems have been implemented on a physical team of robots, and they usually make strong assumptions on the robotic platform such as communication, localization, rich sensory information, and negligible sensor and actuator noise.

- Swarm robotics is emerging as sub-discipline of multi-robot systems. Among the most challenging problems in swarm robotics is the development of appropriate modeling methodologies that allow for the design of the individual platform for achieving a desired collective behavior of the swarm.

- Inspection of jet turbines is a challenging application for multi-robot coverage. The shielded and narrow structure of the turbine imposes drastic constraints on the size and the communication abilities of a potential robotic platform. Real-world systems for the inspection of engineered structures, so far, are semi-autonomous single-unit systems. The nature of the task, however, strongly advocates using multi-robot systems.

CHAPTER 3

Case Study and Experimental Setup

This chapter describes the experimental setup, which aims at modeling part of the challenges that would arise in a turbine inspection task using a miniature robotic swarm (Krasowski, Greer & Oberle 2002). The challenges that need to be overcome for enabling such a scenario are among others miniaturization, coordination despite limited communication and localization, locomotion, and human-swarm interfaces with appropriate expert systems. The experimental setup in this dissertation focuses on the miniaturization and multi-robot coordination aspect. As the motivating case study is concerned with turbine inspection, experiments are conducted in an environment with regularly spaced objects that mimic the blades of the compressor section in a jet turbine. As all algorithms considered in this dissertation require every element to be covered at least once, the experimental setup can also be interpreted as an instance of a graph coverage problem.

Particular emphasis of our experiments is put on sensor and actuator noise and limited/unreliable communication. The chosen platform *Alice* (Caprari & Siegwart 2005) provides only a limited amount of computational power and memory, rather crude sensors and high wheel-slip (Section 3.1.1). The basic Alice robot without extensions fulfills the assumptions of the reactive coverage algorithms presented in Chapter 5. Improved by a secondary micro-controller (Section 3.1.2) the Alice can execute basic deliberative coverage algorithms without explicit collaboration (Section 6.2). Finally, a radio device allows for the sharing of coverage progress (Section 6.3) and near-optimal partitioning

Module	Energy Consumption
Drive-train on	15mW
Drive-train off	< 3mW
Radio active	60mW
Radio sleep	< 1mW
Camera active	60mW
Camera sleep	15mW

Table 3.1: Energy consumption of selected sub-systems of the inspection platform.

of the environment among the robots (Section 6.4). To provide a common reference for collaboration, a camera module (Section 3.1.3) allows the robots to uniquely identify each blade.

3.1   A Miniature Robotic Platform for Autonomous Inspection

Our inspection system (Figure 3.1, *left*) is based on the Alice miniature robot, developed by Caprari & Siegwart (2005) at the Autonomous System Lab at EPFL, which we extended by networking and color imaging capabilities. The robotic platform, the camera, and the radio module can communicate via an I^2C bus. A flow-chart of the whole system is shown in Figure 3.1, *right*. The energy consumption of selected sub-systems of the inspection platform are summarized in Table 3.1.

3.1.1   Basic Platform

The Alice has a size of 21mm x 21mm x 20mm, and is operated by a PIC 16LF877 microprocessor (4Mhz, 384 byte of RAM, 8kB ROM). It is endowed with 4 IR modules which can serve as very crude proximity sensors (up to 3cm) and local communication devices (up to 6 cm in range), providing communication at around 20 nibbles (data sets of 4bit length) per second, and can also be used for detecting the presence and approximate direction of other robots. Its autonomy with a 40mAh (at 3.6V) NiMH rechargeable battery ranges from 10min[1] to 10h, depending on the actuators and sensors used (refer to Table 3.1 for detailed energy consumption of selected components). The Alice is driven by two watch (stepper) motors in a differential-wheels configuration, allowing for a top-speed of 4cm/s.

---

[1]Although the capacity of the battery would allow for more experimental time, in practice the voltage quickly drops below the operational range under high load

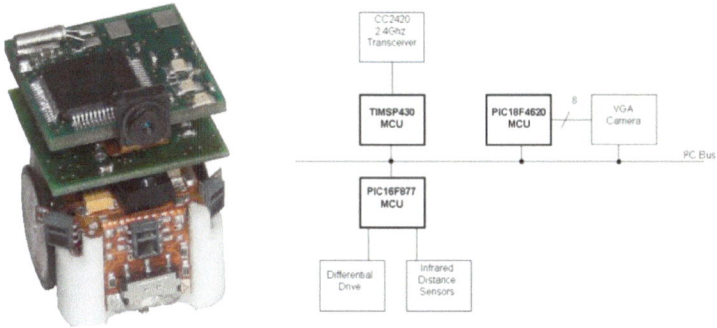

Figure 3.1: The inspection platform (*left*) measuring around 2cm x 2cm x 3cm, endowed with 2 watch motors for differential drive, a 2.4GHz ZigBee- compliant wireless radio, a VGA camera, and three micro-controllers connected by an I^{2}C two-wire bus. Block-Diagram (*right*).

3.1.2   Communication Module

To improve computational and communication capabilities for ad-hoc networking among the robotic swarm and eventually transmission of recorded data to a base station, we developed a 21x21mm^2 extension board that implements functionality identical with the Telos MoteIV mote (Polastre, Szewczyk & Culler 2005), see also Figure 1.3, but has been extended by I^{2}C connectivity (Cianci, Raemy, Pugh & Martinoli 2006). The extension module is operated by a Texas Instruments MSP430 micro-controller (8MHz, 2kB RAM, 60kB ROM), and is endowed with a TI (former ChipCon) CC2420 radio (ZigBee-ready) and 4MByte Flash-Memory. Conveniently, the extension module can be programmed in TinyOS[2] which provides a growing number of ready-to-use libraries for different purposes. The power consumption of the TI micro-controller is extremely small (around 5mW) when compared to that of the radio chip (around 55mW). Although the effective transmission power can be adjusted down by a factor two, the power required for reception is constant.

---

[2]http://www.tinyos.net

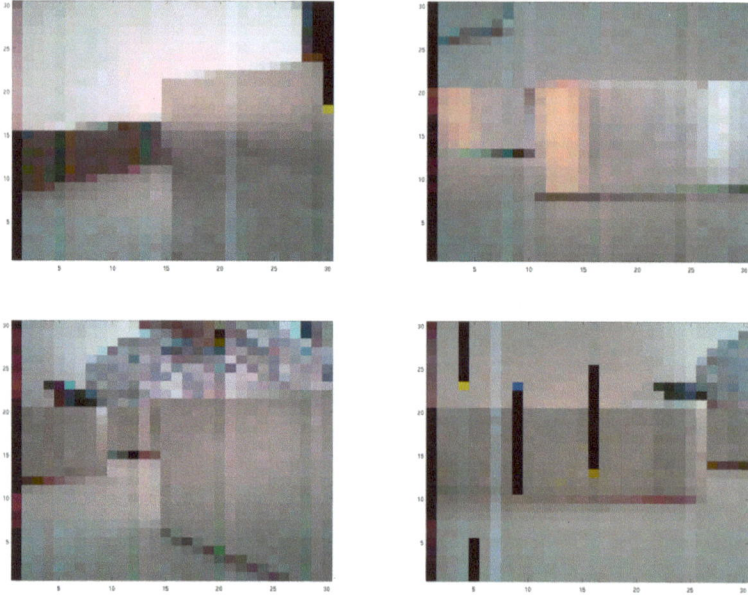

Figure 3.2: Pictures (30x30 pixels) taken by the on-board camera and transmitted over the radio. Vertical black stripes (bottom right) are due to packet loss. The arena boundary (painted in black) can be seen in the top left picture, in the bottom row the experimenter's upper part of the body is visible in the background.

3.1.3   Camera Module

For inspection and localization, we have developed a camera module endowed with a PixelPlus Po3030k VGA miniature camera that is down-sampled at 30x30 pixels in RGB color (RGB-565 coding scheme, 2 bytes per pixel). Using a PIC18LF4620 at 16MHz with 4kB RAM (operational at voltage as low as 2.8V), the Alice is able to uniquely identify color markers in the environment.

Also, the camera module allows the robot to take pictures at a rate of around 2Hz. A series of sample images are depicted in Figure 3.2. Images (1800 byte each) are transmitted over the radio in 72 packets of 25 bytes, where 24 bytes are image data and 1 byte is used for indexing a packet. This allows for partly reconstructing an image in case of packet loss (vertical black stripes in Figure 3.2).

The camera is clocked directly by the micro-controller (division of 32) and its power

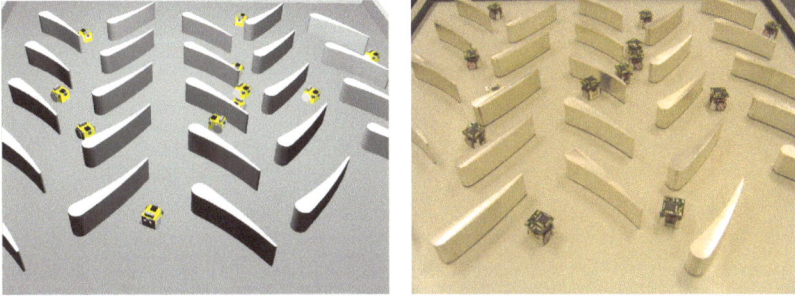

Figure 3.3: *Left*: Overview of the turbine set-up in the realistic simulator. *Right*: Overview of the real-robot set-up. The arena (60x65cm) is equipped with 25 blades and up to 30 miniature robots, mimicking the inspection of a jet turbine's interior.

consumption is independent from the operating frequency. As the power-consumption of the micro-controller is small compared to that of the camera (around a factor three at 16MHz and around a factor of two at 32MHz), running the system at 32MHz might yield considerable savings due to faster image acquisition. However, operation of the micro-controller at 32MHz requires 3.5V supply voltage, which comes close to the nominal voltage of the battery in use and leads to unpredictable results due to occasional brown-out of the micro-controller after very little use of the freshly charged battery. Also, the additional current being drawn leads to higher peak values, which again tend to lower the effective available voltage from the battery.

3.2   Turbine Mock-Up Environment

The swarm of *Alice II* robots is operating in a bounded arena of 60cm x 65cm that is equipped with 25 "blades" (Figure 3.3). The blades are arranged in a regular pattern that is inspired by the compressor section of a turbine's interior. The blades have a length of 11cm, and the ends have a radius of 2mm and 10mm. In this dissertation, the sharp end is referred to as the "blade's tip". The blades are arranged so that two robots that follow the boundary of two different elements do not collide.

We distinguish two scenarios: without localization (Figure 3.3, Chapter 5) and with localization (Figure 3.4, Chapter 6). In scenarios that allow for localization, the upper part of the blades are equipped with a unique color marker that consists of three colored horizontal bars. Presence or absence of the 3 color channels (red, green, and blue) is used to encode 3 bits per color. Using the middle gray bar as reference (all channels at

50%) allows us to encode 64 different codes of which we are using 25 to identify each blade (see for details Rutishauser 2007).

When color markers are absent, robots distinguish between blades and the arena boundaries, exploiting the fact that the outer walls of the arena are painted in black, a color that hardly reflects the infrared light emitted by the robots' distance sensors. When color markers are available, walls simply do not provide an id, which distinguishes them from blades.

In order to support the energy budget of the Alice robot, the environment is also equipped with a static communication module with external power supply that serves as repeater for intra-robot communication and as base station. By this, a robot can turn its radio only on when it requires new information, e.g. prior to a planning step.

3.3   Realistic Simulation

The scenario described in Section 3.2 has been faithfully implemented in the realistic simulator *Webots* (Michel 2004) using the CAD models used for manufacturing the Alice robot and the turbine blades (Figure 3.3, left). Sensor characteristics (aperture, range, and non-linear transfer functions) and the wheel-slip using measurements on the real platform, have been carefully calibrated using real robots (Bronée 2005).

3.4   Performance Evaluation

The setup is monitored from above using a Unibrain Fire-I 400 IEEE1394 (Firewire) camera. The camera provides monochrome images at 30Hz and a resolution of 640x480 pixels. Videos can be recorded to disk and post-processed using the open-source soft-ware *SwisTrack*[3] that has been developed in the course of this thesis (Correll, Sempo, de Meneses, Halloy, Deneubourg & Martinoli 2006).

The robot trajectories will eventually enclose each blade which indicates complete circumnavigation by at least one robot. Using a flood-fill algorithm starting in the center of each blade, it can be tested whether the area of the contour enclosing a blade is below a certain minimal size and therefore assess whether a given blade has been completely inspected. In Webots, trajectories are discretized using a 640 by 480 point grid, which allows for using the same algorithm for measuring coverage progress.

For algorithms requiring communication, performance is also monitored by overhear-ing intra-robot communication using a base station. Due to the potential failure of the

---

[3]http://swistrack.sourceforge.net

Figure 3.4: The experimental setup endowed with colored markers that allow for unique identification of a blade using the on-board camera.

localization algorithm, only overhead vision can be considered as ground truth data, however.

3.5   Discussion

Although the experimental setup of this thesis is limited to the turbine inspection scenario, it has a series of properties that allow for extrapolating experimental results to more general distributed coverage scenarios. For instance, the regular structure of the environment allows for the study of the influence of environmental templates on the spatial distribution of self-organized algorithms (Chapter 5). Also, the algorithms presented in this thesis (Section 6.3 and 6.4), which rely on color markers for global localization, do not take advantage of the regularity of the environment and thus are representative for distributed coverage of arbitrary cellular decompositions in which global localization is available.

Although the NiMH batteries in use have a capacity of 40mAh, effectively available power is dependent on the battery's age. New batteries allow for drawing current up to 120mA for a short time before voltage drops below 3V. These characteristics deplete fast, and after a few charging cycles only a few batteries are able to provide enough current for a parallel operation of the camera and communication module (current consumption around 45mA at 3V), and only if they are fully charged. Although all algorithms considered in this thesis never use communication and camera at the same time, experiments involving any of these devices become unreliable after around 10 to 20 min of operation, due to the potential "brown-out" of the micro-controllers involved. Nevertheless, the resulting heterogeneity of the swarm motivates us for further research on adaptive policies for task decomposition and allocation.

Chapter Summary

- A miniature robotic platform based on the *Alice* platform and comprising ZigBee-ready communication and 30x30 pixels color vision on a footprint smaller than 2cm x 2cm x 2cm has been developed.

- The experiments are conducted in a 60 cm x 65 cm large arena that is a mock-up of the compressor section of a jet turbine engine and consists of 25 "blades". For algorithms requiring localization, each blade is tagged with a unique colored bar-code that can be decoded using on-board hardware.

- The experimental setup allows for studying multi-robot coordination, subject to extreme miniaturization constraints with sensor and actuator noise and unreliable communication.

■ The performance is evaluated using an overhead camera system and the open-source tracking software *SwisTrack*.

CHAPTER 4

Probabilistic Modeling and Identification of Multi-Robot Systems

Models that capture aspects of a multi-robot system and its dynamics are an important design tool. System identification (Johansson 1993, Ljung 1999) is the process of deriving a mathematical model that describes the behavior of a (dynamical) system, and identifying its model parameters from observed data or from a priori available information on the system. Such refined models are then able to provide not only qualitative but also quantitative predictions of the system performance, which allows for exploring different design choices. The design process is usually iterative, i.e. insight gained from models with lower abstraction levels lead to models of improved quality and vice versa. This process is illustrated in Figure 4.1.

This chapter addresses modeling at the macroscopic level and exact microscopic representations of the macroscopic level, i.e. microscopic models that sample possible state space trajectories by simulation without adding additional complexity to the model or changing the modeling assumptions.

One possible way to model swarm-robotic systems is to look at the population dynamics of the swarm (the number of robots in a certain state). This approach (see for instance Sugawara, Sano, Yoshihara & Abe 1998, Martinoli et al. 1999, Lerman & Galstyan 2002) has lead to good quantitative agreements between reality and the prediction of the models (Martinoli et al. 2004, Agassounon et al. 2004). Such models are

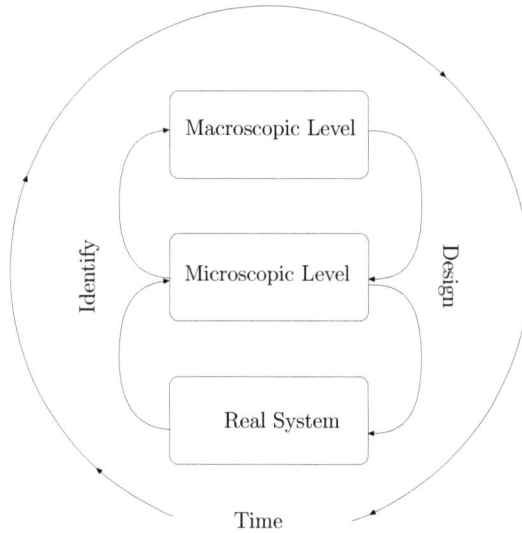

Figure 4.1: The modeling and design process takes place at multiple abstraction levels: the real system, the microscopic level (e.g., agent based models, DES simulation, realistic simulation) and the macroscopic level (e.g., difference equations). The observations of realizations of lower abstraction levels help improve the quality of model predictions at higher abstraction levels. In turn the exploration of design choices at a higher abstraction level motivates design choices (e.g., controllers, behavior).

reasonable for fully reactive systems, where the amount of memory is limited and deliberation is essentially absent and thus leads to a tractable number of states for the robot behavior and the environment.

For robotic systems whose state space consists of the internal representation of the environment and the environment itself, e.g., due to extensive usage of memory for planning among the agents, the state space needs to be modeled by some suitable abstract representation, e.g. a graph that represents the topology of the environment. The probabilistic elements of the robotic sub-system and the environment can then be taken into account by means of their average likelihood or their distribution and explicit simulation, or by formally enumerating every possible constellation and possible state transitions of the system.

This chapter introduces concepts that are important for modeling the distributed boundary coverage case study in this thesis. Models are derived based on probabilistic population dynamic models (Section 4.1) and concepts are illustrated using a series of fully reactive systems and reactive systems with a limited amount of memory (Section 4.2). Section 4.3 then shows how this approach can be be extended for deliberative algorithms subject to noise. Parameter identification based on experimental data (Section 4.4) then allows for improving quantitative prediction of the models.

4.1   Preliminaries

This section introduces the notion of state and proposes some rough guidelines for selecting an appropriate state-space granularity. It then introduces the concept of Master equations that track changes in the probability of a system being in a particular state, and Rate equations that track the number of agents in a particular state.

4.1.1   State-space Granularity

The state X of a multi-robot system is uniquely defined by the Cartesian product of the states $R_i = \omega$ for $i \in \{1, \ldots, N_0\}$ of the robotic system described by N_0 random variables and the states of the environment $E_j = \psi$ with $j \in \{1, \ldots, M_0\}$ described by M_0 random variables. Thus

$$X = R_1 \times \ldots \times R_{N_0} \times E_1 \times \ldots \times E_{M_0} \qquad (4.1)$$

describes the state of a multi-robot system and the environment.

The random variables can take values out of a set $\omega \in \Omega$ and $\psi \in \Psi$ that correspond to possible robot and environmental states, respectively.

When describing and simulating a system, the system state definition (4.1) might be unhandy and there are more suitable representations, e.g. graphs or Petri-nets.

In practice, the definition of the system state is depending on the metric of interest. For the robotic system Ω can usually be derived directly from the Finite State Machine (FSM) of the robot controller. For the environment, the choice of Ψ depends on the metric of interest, e.g. a door can be open or closed, a light can be red, green, or orange. Choosing the appropriate level of detail for a model has to be carefully evaluated and potentially adapted during the modeling process (see also the discussion of Lerman et al. 2005).

In this thesis Ω and Ψ are always discrete sets, leading to a non-infinitesimal state space.

4.1.2   Master Equations, Rate Equations, and Markov chains

For the probability of a random variable to have the value ω, one can write the following time-discrete Master equation

$$p_\omega(kT + T) = p_\omega(k) + \sum_{\omega' \in \Omega \setminus \omega} (p_{\omega'\omega}(kT + T)p_{\omega'}(k) - p_{\omega\omega'}(kT)p_\omega(kT)) \qquad (4.2)$$

where $p_{\omega'\omega}(kT + T)$ is the conditional probability that the system will be in state ω at time $kT + T$ when in state ω' at time kT. Notice, that $p_{\omega'\omega}(kT + T)$ can be also understand as the transition probability of state ω to ω' of a *Markov chain* with states Ω (Cassandras 1993). T is the time discretization of the system and k is indexing the time-steps. For brevity, T is omitted in the notation in the remainder of this dissertation.

Using $p_x(k)$ and the total number of individuals N_0, $p_x(k)$ can also be interpreted as the fraction of robots in state x and $N_0 p_x(k)$ yields the the expected number of robots in state x. For brevity, it is written

$$N_x(k) = N_0 p_x(k) \qquad (4.3)$$

Using this notation, (4.2) can be written as

$$N_x(k + 1) = N_x(k) + \sum_i (p_{ix}(k + 1)N_i(k) - p_{xi}(k)N_x(k)) \qquad (4.4)$$

which is a discrete-time Rate equation. The conditional probabilities can be either constant, time-varying or a function of other states of the system. In the latter case, (4.2) and (4.4) become *non-linear*. In this case, the system can be understood as a Markov chain with time varying state transition probability matrix (time-heterogeneous Markov chain).

4.1.3   Continuous Time vs. Discrete Time Models

Both (4.2) and (4.4) can also be formulated in continuous time (see for instance Lerman et al. 2005) leading to a set of differential equations.

In this thesis, time-discrete models are used whenever models are given by compact, macroscopic representations. The time discretization of the model is chosen so that T is reasonable small when compared with the time granularity of the state durations observed in the real system. At the microscopic level, simulation of a series of synchronous, stochastic automata then corresponds to an exact representation of such a system[1] (Martinoli et al. 2004).

Time-continuous models are applied in this thesis when the asynchronicity of the system needs to be explicitly taken into account for calculating the performance metric. In this case, analytical solutions are usually untractable but for special cases (see Example 5, Section 4.3.2), and performance is analyzed by DES simulation.

4.1.4   Modeling Assumptions

The models require that the environment and the multi-agent system can be described as Probabilistic Finite State Machine (PFSM); the state granularity of the description is established as a function of the metric of interest, and the desired level of detail (Section 4.1.1). Also, the models require the robotic system and the environment to be *semi-Markovian*, i.e. the system's future state is only a function of the current state and the time spent in this state (Cassandras 1993) so that the probability for the system to be in a certain state can be written as defined in Section 4.1.2). Finally, unless the robot position is explicitly encoded in the state space, all models assume that the spatial distribution of the agents in the environment is uniform (well-mixed), and that the environment is bounded. In turn, only non-spatial metrics are considered. For example, a model tracks the average number of inspected elements, but not whether a specific element is covered or not.

4.1.5   Model Parameters

The number of required parameters depends on the abstraction level and usually decreases with an increasing abstraction level. Parameters for realistic simulation are either a priori known from a robot's data-sheet or can be easily measured using individual robots (e.g., the non-linearity of a robot's distance sensor, the robot's speed, or wheel-slip). In probabilistic models, however, all parameters need to be abstracted into state durations and transition probabilities between states. In the formalism described

---

[1]For differential equations, the Gillespie algorithm is an exact representation

here, state duration and transition probabilities are explicitly defined by the likelihood of interaction between robots, and between robots and the environment; state durations define the time an interaction lasts.

State durations and Time Discretization

State durations are either of probabilistic nature — for instance the time to avoid an obstacle, or the time to manipulate an object — or deterministic and hard-coded into the robot's controller.

State durations are measured in time-steps, which correspond to the time-discretization of the macroscopic model and its length T can be arbitrarily chosen. For facilitating computation, it makes sense to choose the smallest common factor of all time delays that are necessary to describe the metric of interest with sufficient precision (see for further details Martinoli et al. 2004, Correll & Martinoli 2004b).

States with a fixed duration $T > 1$ are referred to as *delay states* and consist of a series of T states, which are left with probability 1 after 1 time-step.

Encountering Probabilities

The encountering probability for an object is the probability that a robot encounters this object within a time-step when moving through the environment. A constant encountering probability that is independent of the position of the robot requires (1) robots to be uniformly distributed in the environment, and (2) that the objects have sufficient distance so that a robot can only detect one object at a time. While spatial uniformity can be assumed when robots show trajectories that correspond to a random walk (Rudnick & Gaspari 2004, Prorok 2006), linear super-position of encountering probabilities is not always possible, especially when the density of objects and robots is high.

Sensor and Actuator Noise

Whereas encountering probabilities define the transition probabilities in reactive systems, in deliberative systems transition probabilities between two states are given by the control policy. Transition probabilities are either 1 or 0 if the system would be deterministic and correspond to the likelihood to succeed or not succeed, respectively, a desired action under the influence of randomness or sensor and actuator noise. Whereas randomness at the controller level can be directly imported into the model, the probability of failure to execute a particular state transition due to sensor and actuator noise needs to be calibrated on the real system or using realistic simulation.

4.1.6 Transient, Recurrent, Periodic and Absorbing States

States can be classified in transient, recurrent, periodic and absorbing states, which will be seen to be important for steady-state analysis. This classification is customary in Markov chain analysis (see for instance Cassandras 1993) and are also useful for describing non-linear models such as those used in this chapter.

A state is *transient* if there is a non-zero probability that the system never enters this state again once it has been in it. Transient states are typically used for modeling irreversible processes such as coverage progress or auxiliary states within delay states.

A state is *recurrent* if it is not transient, i.e. the system will reach it again with probability 1 sooner or later. Recurrent states are typically used for modeling the default state of a system, such as being idle or searching, e.g.

A state ω has period k if any return to state ω must occur in some multiple of k time steps and k is the largest number with this property (it is *k-periodic*). If $k = 1$, then the state is said to be *aperiodic*.

A state is *absorbing* if it is impossible to leave this state once it has been entered. Absorbing states are typically used for modeling terminal conditions such as robot failure or task completion.

4.1.7 Steady-State Analysis

Whether a system reaches a steady or stationary state or not is important when investigating convergence, e.g. of a coverage problem. There are two important theorems that will be provided without proof (see, e.g., Asmussen 1987) for irreducible and reducible Markov chains that will be used in this thesis (see also Section 4.1.2). A Markov chain is irreducible if all states can be reached from any other state. It is reducible if this is not the case, e.g., when there exist absorbing states.

Theorem 4.1.1. *In an irreducible Markov chain consisting with finite state space $\omega \in \Omega$ and no periodic states, a unique stationary state probability exists such that*

$$\lim_{k \to \infty} p_\omega(k) = \frac{1}{M_\omega} \tag{4.5}$$

where M_ω is the mean recurrence time of state ω and independent from the initial conditions.

Theorem 4.1.2. *In a reducible Markov chain, the chain eventually enters a set of irreducible states with a stationary distribution according to Theorem 4.1.1. If the irreducible sub-set of the chain contains an absorbing state this will be reached with probability 1.*

Numerical values for the steady-state are straightforward to obtain for linear systems and are given by the Eigenvector with Eigenvalue 1 of the state transition probability matrix of the system. Alternatively, difference equations can be explicitly solved in the time or the frequency domain (Martinoli et al. 2004). Finally, for delay states with length T, the stationary distribution can be obtained by integrating all transition probabilities leading to this state from time-step 0 to T.

For non-linear systems only few instances can be solved analytically and numerical values have to be obtained by numerically integrating the underlying difference equations. Alternatively, the stationary distribution can be estimated by sampling realizations of the system in simulation.

4.2   Probabilistic Models for Reactive Systems with Limited Memory

This section illustrates concepts introduced in Section 4.1 by modeling a series of multi-robot systems with reactive robot controllers. These examples have been selected as they can be considered generic building blocks for modeling complex interactions among a team of robots and the environment. For each controller, deterministic and probabilistic state durations are considered.

4.2.1   Example 1: Collision Avoidance

As a first example, consider a scenario where a swarm of N_0 robots is moving in an empty, enclosed arena. The robots behavior is defined by a FSM having only two states, $\Omega = \{\texttt{Search}, \texttt{Avoidance}\}$. Assuming a uniform distribution for the robots in the environment and linear super-position of encountering probabilities (see Sections 4.1.4 and 4.1.5), potential collisions with other robots are summarized by the encountering probability $p_R = p_r(N_0 - 1)$ and the time T_a that a collision takes, where p_r is the encountering probability of a single robot. For simplicity, collision with walls (an additional state) are ignored. As the interaction of the robot with other robots is considered to be probabilistic, the i-th robot's state is given by the random variable $R_i = \omega \in \Omega$.

Probabilistic Delay

The time spent in a state is probabilistic, and the probability to enter $\texttt{Search}$ from $\texttt{Avoidance}$, i.e. finish avoidance, is given by $P\left(R_i(k+1) = \{\texttt{Search}\}|R_i(k) = \{\texttt{Avoidance}\}\right) = \frac{1}{T_a}$. This assumption is reasonable for robots with proximal controllers that require an average time of T_a for avoiding an obstacle. Proximal controllers are reactive, and the wheel-speeds are a direct function of some sensor measurements (see also Braitenberg 1986, Arkin 2000).

The PFSM for this system is depicted in Figure 4.4, a. Using (4.4) the number of robots in states Search and Avoidance at time k is then given by

$$N_s(k+1) = N_s(k) - p_R N_s(k) + \frac{1}{T_a} N_a(k) \qquad (4.6)$$

$$N_a(k+1) = N_0 - N_s(k+1) \qquad (4.7)$$

and by the fact that the total number of robots is constant in the enclosed arena. Possible initial conditions are $N_s(0) = N_0$ and $N_a(0) = 0$, i.e. all robots are in search mode at the beginning.

Given a probability $\frac{1}{T_a}$ to leave a state, the average time spent in Avoidance is the expected value $E[T_a] = T_a$ which is calculated by summing over all possible state durations $j = [0; \infty]$ multiplied by its probability

$$E[T_a] = \sum_{j=0}^{\infty} \left(1 - \frac{1}{T_a}\right)^{j-1} \frac{1}{T_a} j \qquad (4.8)$$

$$= \frac{1}{T_a - 1} \sum_{j=0}^{\infty} \left(1 - \frac{1}{T_a}\right)^{j} j$$

$$= \frac{1}{T_a - 1} \frac{(1 - \frac{1}{T_a})}{(1 - (1 - \frac{1}{T_a}))^2}$$

$$= T_a$$

Thus, a robot will spend T_a time steps in Avoidance on *average*.

The system described above has been simulated on two different microscopic abstraction levels, Webots, and synchronous agent-based simulation. In Webots a robot was randomly walking in a small quadratic arena with a static obstacle in the middle. Obstacles were avoided using a proximal (reactive) controller. Agent-based simulation was implemented in Matlab by simulating the PFSM from Figure 4.4a. Both times, the state duration of the avoidance state has been measured from the perspective of the agent (egocentric measurement) and of a supervisor (allocentric measurement). For egocentric measurements, a robot measures the time it encounters an obstacle using its on-board sensors. For allocentric measurements, a supervisor measures the time a robot is less than the maximal range of its distance sensors apart from any obstacle.

Results are depicted in Figure 4.2. As desired, the average time spent in Avoidance at the synchronous, agent-based level shows a geometric distribution (Figure 4.2, *left*). Although providing the same average value ($T_a = 0.75$s) than egocentric measurements obtained in Webots, its distribution (Figure 4.2, *middle*) is not quite geometric, yet not

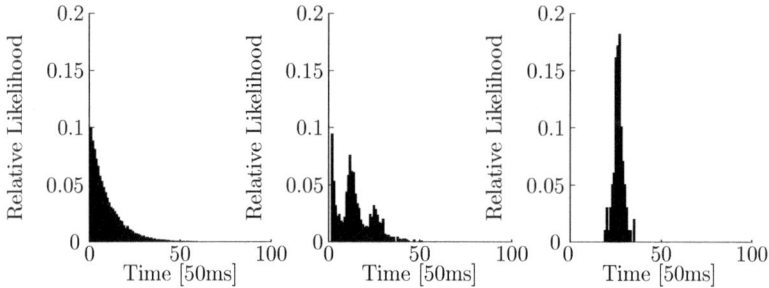

Figure 4.2: Relative likelihood of the time spend in avoidance in a collision avoidance experiment using a proximal controller. Egocentric measurements in synchronous, agent-based simulation (*left*), egocentric measurements in Webots (*middle*), and allocentric measurements in Webots (*right*).

Gaussian. Finally, Figure 4.2, *right*, shows allocentric measurements for T_a obtained in Webots and result in $T_a \approx 1.32$s. This is due to the fact that when observing the robot according to a specific criterion (here: distance to the obstacle less than the sensor range of the robot), a supervisor cannot distinguish between a current or a resolved collision, i.e. the robot is moving away from the obstacle.

These results illustrate well the approximative character of the proposed model and the requirement for precise definition of a system's performance metric.

Although the slight differences in the two different egocentric models will yield the same result when integrated in a linear model (due to the identical mean), the underlying distributions might become important in a non-linear model. In this case, more faithful prediction might be achieved by using a suitable parameterization or to sample directly from the observed distribution.

Whereas the probabilistic is a reasonable approximation of the egocentric measurements on the robot, the allocentric measurements would have been better captured by a deterministic, constant delay (compare Figure 4.2, *right*).

Deterministic Delay

The time spent in Avoidance is now deterministic and has length T_a time-steps, e.g. because a robot always turns around 180 degree upon collision instead of reactively avoiding the obstacle. A robot's controller will then return into state Search with probability $P(R_i(k+1) = \{\text{Search}\}|R_i(k) = \{\text{Avoidance}\}) = 1$ after T_a steps, which requires memory and a counter. As the system's future state depends not only on its current state but on the time spent in this state, the system is a generalized semi-Markov

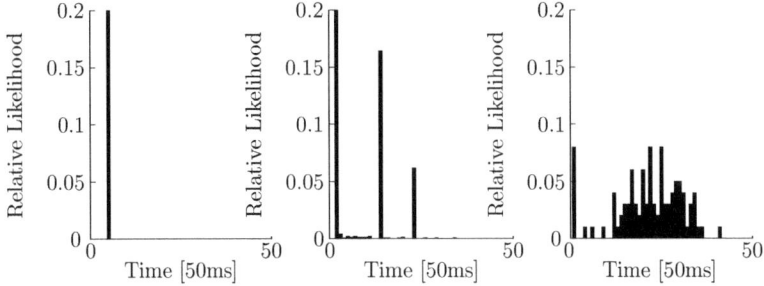

Figure 4.3: Relative likelihood of the time spend in avoidance in a collision avoidance experiment using a distal controller. Egocentric measurements in synchronous, agent-based simulation (*left*), egocentric measurements in Webots (*middle*), and allocentric measurements in *Webots*.

process of order T_a.

The system dynamics are described by

$$N_s(k + 1) = N_s(k) - p_R N_s(k) + p_R N_s(k - T_a) \qquad (4.9)$$

$$N_a(k + 1) = N_0 - N_s(k + 1) \qquad (4.10)$$

In words, the number of robots in Search is decreased by those entering Avoidance with probability p_R, and increased by those that entered Avoidance before T_a time-steps. The number of robots is zero for negative iterations, i.e.

$$N_s(k) = N_a(k) = 0 \; \forall \; k < 0 \qquad (4.11)$$

The PFSM for this system is given in Figure 4.4, *b*. Equation 4.9 can be derived by reformulating the system as strictly Markovian process consisting of T_a transient states of duration 1 and substituting the resulting equations into each other.

As for the probabilistic delay, the system was implemented at two different abstraction levels. This time, the simulated robot in Webots uses a distal controller and turns 180 degree upon encountering an obstacle. In Matlab, the PFSM from Figure 4.4b has been simulated. The results are shown in Figure 4.3.

Due to the constant time required for doing a U-turn, the system is simulated using constant delay for Avoidance (Figure 4.3, *left*). Egocentric measurements in Webots reveal however, that the collision is only avoided in around 75% of the cases using this strategy, and requires two or more U-turns otherwise (Figure 4.3, *middle*). Both times, the average is $T_a = 0.27$s. Figure 4.3, *right* shows allocentric measurements for the collision duration based on position measurements of the robot. As allocentric

a)

b)

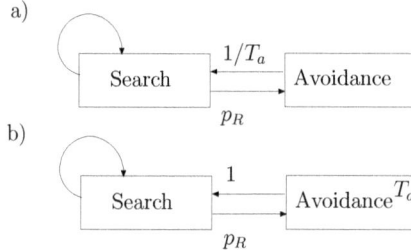

Figure 4.4: PFSM for a robot with two states. Probabilistic delay (a), and deterministic delay (b).

measurements do not distinguish between robots before and after the U-turn, the average avoidance duration is measured to $T_a \approx 1.07$s.

The results show that the accuracy of model prediction depends on careful identification of the system dynamics and exact definition of the performance metric. In this example the egocentric perception of the Avoidance state might have been better modeled using a two-state sub-chain consisting of one deterministic and one probabilistic delay for modeling the U-turn and its success probability, respectively. Alternatively, the empirical distribution of the egocentric measurements could be parameterized or sampled from measurements. Similarly, using the distribution itself in the model would allow for exact simulation of the allocentric perception of the Avoidance state.

Steady-State Analysis

Both (4.6) and (4.9) show the same behavior at steady state (Section 4.1.7), which is given by

$$\begin{pmatrix} N_s^* \\ N_a^* \end{pmatrix} = \lim_{k \to \infty} \begin{pmatrix} N_s(k) \\ N_a(k) \end{pmatrix} = \begin{pmatrix} \frac{N_0}{1 + p_R T_a} \\ \frac{N_0 p_R T_a}{1 + p_R T_a} \end{pmatrix} \tag{4.12}$$

The steady-state solution for both systems assumes p_R and T_a to be constant. Although analytical solutions are also possible for linear systems where p_R and T_a are probabilistic with certain distributions, simulating the system and sampling from these distributions is a valuable option, in particular when the distributions are not available in parameterized form. The steady-state solutions only yield average values but not the distribution of the performance, which can also be obtained using simulation.

4.2.2   Example 2: A Basic Environmental Model

As second example, consider a simple inspection task where a number of S sites in the environment needs to be visited by at least one robot. Every site has the encountering probability p_s to be visited by an individual robot every time step. Site S_i can thus be in two states, unvisited and visited, i.e. $\Psi = \{\texttt{unvisited}, \texttt{visited}\}$ and $p_{i,u}$, $p_{i,v}$ the probability of site being unvisited or visited, respectively. As a visit is irreversible, $\texttt{visited}$ are absorbing states. For a large number of sites and $N_s(k)$ searching robots, the likelihood for a site i to be visited is then given by the following difference equation:

$$p_{i,u}(k+1) \quad = p_{i,u}(k) - p_s N_s(k) p_{i,u}(k) \tag{4.13}$$

$$p_{i,v}(k+1) \quad = 1 - p_{i,u}(k+1) \tag{4.14}$$

with $p_{i,u}(0) = 1$ and $p_{i,v}(0) = 0$.

The expected number of unvisited sites is given by

$$U(k) = \sum_{i=0}^{S-1} p_{i,u}(k) \tag{4.15}$$

Substituting (4.13) into (4.15) yields

$$U(k+1) \qquad = \sum_{i=0}^{S-1} p_{i,u}(k) - p_s N_s(k) \sum_{i=0}^{S-1} p_{i,u}(k)$$

$$\Leftrightarrow \quad U(k+1) \qquad = U(k) - p_s N_s(k) U(k) \tag{4.16}$$

and thus summarize S states by a single difference equation. Assuming the number of searching robots being at steady-state N_s^*, we can solve (4.16) to

$$U(k) = S(1 - p_s N_s^*)^k \tag{4.17}$$

and $\lim_{k \to \infty} U(k) = 0$ and $\lim_{k \to \infty} V(k) = S$.

4.2.3   Example 3: Collaboration

Consider a system where every robot of a team of N_0 can be in two states $\Omega = \{\texttt{Search}, \texttt{Wait}\}$. Also there are M_0 sites where robots can collaborate. When in $\texttt{Search}$, a robot encounters a site with probability p_s. If the site is empty, i.e. no other robot around, a robot enters $\texttt{Wait}$ and waits for another robot to collaborate or until a time-out expires. If the site is occupied by a waiting robot, robots collaborate and both return to $\texttt{Search}$. For simplicity, we assume collaboration to take place within the current time step, i.e. both robots involved return into $\texttt{Search}$ in the next time-step.

a)

$$1/T_w$$

Search   Wait

$$p_s(M_0 - N_w(k))$$

$$p_s N_s(k)$$

$$p_s(M_0 - N_w(k - T_w))\Gamma(k; k - T_w)$$

b)

Search   Wait T_w

$$p_s(M_0 - N_w(k))$$

$$p_s N_s(k)$$

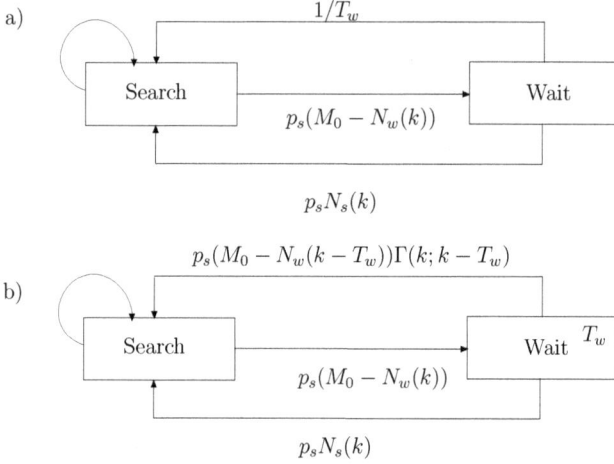

Figure 4.5: PFSM for a collaboration model with two states. Probabilistic delay (a), and deterministic delay (b).

Probabilistic Delay

Robots leave `Wait` with probability $P(R_i(k+1) = \{\texttt{Search}\}|R_i(k) = \{\texttt{Wait}\}) = \frac{1}{T_w} + p_s N_s(k)$, leading to an *average* waiting time of T_w time-steps for robots that do not collaborate. The PFSM of this system is shown in Figure 4.5, *a*, and the system dynamics are given by

$$N_s(k+1) = N_s(k) - p_s N_s(k)(M_0 - N_w(k)) \tag{4.18}$$
$$+ \frac{1}{T_w} N_w(k) + p_s N_s(k) N_w(k)$$
$$N_w(k+1) = N_0 - N_s(k+1) \tag{4.19}$$

with $N_s(0) = N_0$ and $N_w(0) = 0$.

Deterministic Delay

The model with deterministic delay has originally been introduced for modeling the collaborative aspect of the stick-pulling experiment (Martinoli et al. 2004, Lerman, Galstyan, Martinoli & Ijspeert 2001) where robots wait for a fixed amount of time for collaboration. This model has been proven useful also for modeling collaboration among robots in the inspection case study (Section 5.2). Although the deterministic delay involves memory, the model from Martinoli et al. (2004) can be developed based on an equivalent, strictly Markovian system.

After encountering a site, a robot waits at a site (`Wait`) until one of the two events occur: (1) another robot joins the same site or (2) T_w time-steps pass without another robot joining. The state `Wait` can thus be broken down into T_w states $w_0, w_1, \ldots, w_{T_w-1}$, where state w_i corresponds to the $i-th$ time interval a robot spends in `Wait`. It follows $N_w(k) = \sum_{j=0}^{n-1} N_{w_j}(k)$ for the total number of robots in `Wait`. The number of robots in `Search` is then given by

$$N_s(k+1) = N_s(k) \qquad -p_s N_s(k)(M_0 - N_w(k)) \qquad (4.20)$$
$$+p_s N_s(k) \sum_{j=0}^{T_w-2} N_{w_j}(k) + N_{w_{T_w-1}}(k)$$

where the first term corresponds to the number of robots encountering an empty site. The sum over N_{w_0} to $N_{w_{T_w-2}}$ summarizes the robots that collaborate within the time interval $[0; T_w - 2]$, and the last term corresponds to the robots that return after T_w time steps either due to time-out or due to collaborating at time $T_w - 1$. Thus (4.20) can be rewritten as

$$N_s(k+1) = N_s(k) \qquad -p_s N_s(k)(M_0 - N_w(k)) \qquad (4.21)$$
$$+p_s N_s(k)N_w(k) + N_{w_{T_w-1}}(k)(1 - p_s N_s(k)))$$

The number of robots waiting is given by

$$N_{w_0}(k+1) = \qquad p_s N_s(k)(M_0 - N_w(k)) \qquad (4.22)$$
$$\vdots$$
$$N_{w_i}(k+1) = \qquad (1 - p_s N_s(k))N_{w_{i-1}}(k) \qquad (4.23)$$
$$\vdots$$
$$N_{w_{T_w-1}}(k+1) = \qquad (1 - p_s N_s(k))N_{w_{T_w-2}}(k) \qquad (4.24)$$

By inserting (4.23) for $i = 1 \ldots T_w - 2$ into (4.24), we obtain

$$N_{w_{T_w-1}}(k+1) = N_{w_0}(k - (T_w - 1)) \prod_{j=0}^{T_w-2} (1 - p_s N_s(k - j)) \qquad (4.25)$$

and using (4.22)

$$N_{w_{T_w-1}}(k) = p_s N_s(k - T_w)(M_0 - N_w(k - T_w)) \prod_{j=1}^{T_w-1} (1 - p_s N_s(k - j)) \qquad (4.26)$$

By substituting (4.26) in (4.20) we can calculate the number of robots in Search to

$$N_s(k+1) = N_s(k) - p_s N_s(k)(M_0 - N_w(k))$$
$$+ p_s N_s(k) N_w(k)$$
$$+ p_s N_s(k - T_w)(M_0 - N_w(k - T_w)) \prod_{j=0}^{T_w - 1} (1 - p_s N_s(k - j)) \tag{4.27}$$

where $N_w(k) = N_0 - N_s(k)$. As argued by Lerman et al. (2001) and Martinoli et al. (2004) $\prod_{j=0}^{T_w-1}(1 - p_s N_s(k-j))$ can be understand as the fraction of robots that did not collaborate during the time interval T_w. The PFSM summarizing this system is shown in Figure 4.5, *b*.

4.3   Probabilistic Modeling of Deliberative Systems

Deliberative systems are usually well represented by deterministic models. Depending on the domain, analytical insight can be obtained by reformulating the problem into an equivalent one which has been well studied, e.g., from the class of $\mathcal{NP}$-hard problems in Operational Research or multi-player (collaborative) games in Game Theory. Under the influence of sensor and actuator noise, including noisy communication and localization, a theoretical optimal policy might well lead to a sub-optimal outcome, however. In Game Theory this effect is known as "trembling hand perfect equilibrium" (Selten 1975), and takes the possibility into account that the players may choose unintended strategies through a "slip of the hand". Under the influence of noise, a *pure strategy* (every move of the strategy is played with probability 1) becomes *totally mixed strategy* (every possible pure strategy is played with a non-zero, positive probability).

This section develops probabilistic models for basic deliberative algorithms without and with collaboration among agents that are representative for the deliberative coverage algorithms developed in Chapter 6.

4.3.1   Example 4: A Basic Task Allocation Problem

Consider a system with two agents and two tasks A and B. Both tasks are executed within one time-step. However, A is more "attractive" for both agents and will be executed preferably. The information which tasks need to be done and what benefits they yield are available to the robots at any time.

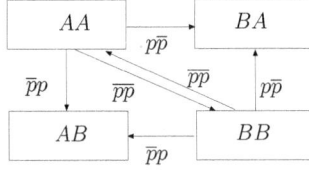

Figure 4.6: PFSM for a deliberative task allocation problem with two robots and two tasks A and B. The states represent the permutations of all possible allocations. Transition probabilities reflect deliberative actions subject to error with probability p. Once both tasks have been accomplished (states AB and BA), both robots stop.

A Basic Task Allocation Problem without Collaboration

A non-collaborative, deterministic, greedy algorithm would now yield the sequence AB for both robots, and thus complete both tasks after two time-steps. Consider now that choosing the task is subject to noise, and an agent would choose the less beneficial task with probability p. In a real system, this can be either due to explicit randomized behavior or due to sensor or actuator noise. In this example, randomness is beneficial, as it allows completing both tasks in one time-step with probability $p\bar{p} + \bar{p}p$, i.e. one of the two robots fails to perform the preferred task A. The system can then be in 4 different states $\Psi = \{AA, BB, AB, BA\}$, i.e. both robots performing task A, both robots performing task B, or one robot task A and the other robot task B (see Figure 4.6). The time-varying probability for being in any of these states is denoted p_{AA}, p_{BB}, p_{AB}, and p_{BA}, and the master equations are

$$p_{AA}(k+1) = p_{AA}(k) - p_{AA}(k)\bar{p}p - p_{AA}(k)p\bar{p} - p_{AA}(k)\overline{pp} + p_{BB}(k)\overline{pp} \qquad (4.28)$$
$$p_{BB}(k+1) = p_{BB}(k) - p_{BB}(k)\bar{p}p - p_{BB}(k)p\bar{p} - p_{BB}(k)\overline{pp} + p_{AA}(k)\overline{pp}$$
$$p_{AB}(k+1) = p_{AB}(k) + p_{BB}(k)\bar{p}p + p_{AA}(k)p\bar{p}$$
$$p_{BA}(k+1) = p_{BA}(k) + p_{AA}(k)\bar{p}p + p_{BB}(k)p\bar{p}$$

with initial conditions $p_{AA}(0) = \overline{pp}$ (both robots pick the right task), $p_{BB}(0) = pp$ (both' robots pick the wrong task) and $p_{AB}(0) = p_{BA}(0) = \bar{p}p$ (one of the robots picks the wrong task).

All possible completion times and their probabilities can then be enumerated as follows

$$T = 1 : p\bar{p} + \bar{p}p \tag{4.29}$$

$$T = 2 : (\overline{pp} + pp)(1 - pp) \tag{4.30}$$

$$T = 3 : (\overline{pp} + pp)pp(1 - pp) \tag{4.31}$$

$$\vdots$$

$$T = k : (\overline{pp} + pp)(pp)^{k-2}(1 - pp) \tag{4.32}$$

Similar as for (4.8) we can calculate the expected value for the completion time T by

$$
\begin{aligned}
E[T] &= p\bar{p} + \bar{p}p + \frac{1}{(pp)^2}\sum_{k=0}^{\infty}(\overline{pp} + pp)(pp)^k(1 - pp)k - \frac{(\overline{pp} + pp)(1 - pp)}{pp} \\
&= p\bar{p} + \bar{p}p + \frac{(\overline{pp} + pp)}{1 - pp} - \frac{(\overline{pp} + pp)(1 - pp)}{pp} \\
&= \frac{(\overline{pp} + pp)(pp - 2) + 2\bar{p}p(pp - 1)}{pp - 1}
\end{aligned}
\tag{4.33}
$$

The above example uses two important assumptions: first, identification of available tasks and their benefit is deterministic, and second, task execution takes exactly one time-step. The problem becomes increasingly difficult when the evaluation of benefits of a task or whether a task needs to be done or not is noisy, or when time needed for a task is also probabilistic. If noise models and execution times are available as average values or distributions, *simulation* of the above system allows for quick insight.

The expected value for the non-collaborative policy for various values of p is shown in Figure 4.8.

A Basic Task Allocation Problem with Collaboration

In this example robots can collaborate and will optimally distribute the tasks among them using a deterministic algorithm. Under absence of noise, the first agent will perform task A and the second agent task B. We again assume that an agent will pick the wrong task with probability p. The benefit of collaboration becomes immediately clear as the chance of completing both tasks within the first time-step is given by $\bar{p}\bar{p} + pp$, i.e. both agents achieve to follow the optimal policy or both agents execute the wrong task at the same time, which is higher than $p\bar{p} + \bar{p}p$ for $p < 0.5$. In case of failure, both agents will be aware of the missing task (again assuming this perception is deterministic), and both plan to execute it. This system can again be represented by four states and associated master equations with initial conditions $p_{AA}(0) = \bar{p}\bar{p}$, $p_{BB}(0) = pp$, $p_{AB}(0) = \overline{p}\overline{p}$, and $p_{BA}(0) = pp$. The system dynamics are again given by (4.28).

For calculating the expected value of the time to completion, we can enumerate all possible outcomes of an experiment and their probabilities as follows

$$T = 1 : \overline{pp} + pp \tag{4.34}$$

$$T = 2 : (p\overline{p} + \overline{p}p)(1 - pp) \tag{4.35}$$

$$T = 3 : (p\overline{p} + \overline{p}p)pp(1 - pp) \tag{4.36}$$

$$\vdots$$

$$T = k : (p\overline{p} + \overline{p}p)(pp)^{k-2}(1 - pp) \tag{4.37}$$

and calculate the expected value for the completion time T by

$$
\begin{aligned}
E[T] &= \overline{pp} + pp + \frac{1}{(pp)^2} \sum_{k=0}^{\infty} (p\overline{p} + \overline{p}p)(pp)^k (1 - pp)k - \frac{(p\overline{p} + \overline{p}p)(1 - pp)}{pp} \\
&= \overline{pp} + pp + \frac{1}{pp} \frac{(p\overline{p} + \overline{p}p)}{1 - pp} - \frac{(p\overline{p} + \overline{p}p)(1 - pp)}{pp} \\
&= \overline{pp} + pp + \frac{2\overline{p}p(pp - 2)}{pp - 1} = 1.6144
\end{aligned}
\tag{4.38}
$$

As expected, the collaborative policy has a lower expected value for completion time and is thus more robust to failure. Again, analytical solutions as those obtained here are unfeasible in practice as the state-space becomes quickly intractable with the number of states (see also Figure 4.7 for an example with four tasks and 2 robots) and the level of additional uncertainty, e.g. the likelihood of successful communication, successful task evaluation, or successful task completion. Simulating such a system using microscopic models as described in (Martinoli et al. 2004) or DES simulators are then as exact representations as for less complex Markov chains like those for the reactive controllers above.

The expected value for the collaborative policy for various values of p is shown in Figure 4.8. For $p = 50\%$ both policies are equivalent. Although the non-collaborative policy benefits from noise up to the critical level of 50%, the collaborative policy decays more gracefully for large noise levels and is always better.

4.3.2   Example 5: A Basic Coverage Problem

Consider a Hamiltonian cycle $\mathcal{H}$ of length S, which covers S vertices denoted s_i that correspond to a cellular decomposition of an environment. For some grids, such a Hamiltonian cycle can be constructed in linear time using Gabriely & Rimon's (2001) STC algorithm, see Figure 4.9 for an example. Assuming N robots, and S to be a multiple of N, the Hamiltonian cycle can be equally partitioned into N parts, which can be covered independently by a team of robots (Hazon & Kaminka 2005).

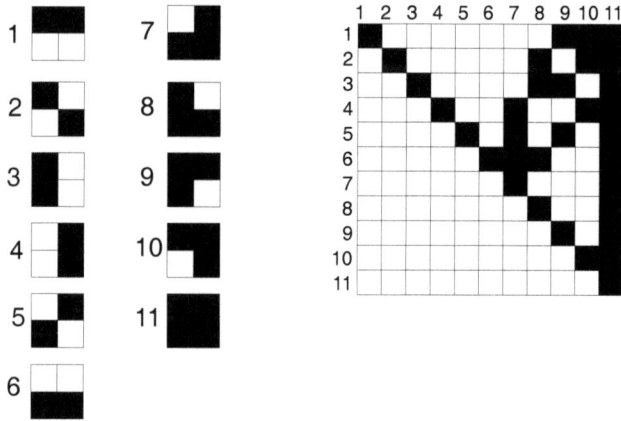

Figure 4.7: A 2x2 grid needs to be covered by two robots. The state space is shown to the left with state 1–6 being possible initial conditions. The state transition probability matrix is shown to the right.

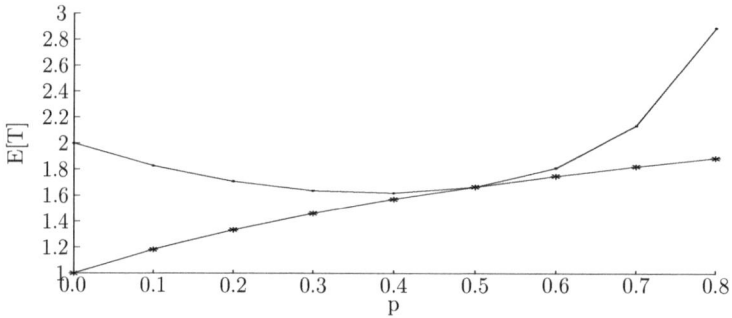

Figure 4.8: Expected value for time to completing two tasks by two robots using a non-collaborative, greedy policy (–·–) and a collaborative policy (–*–). Both policies are subject to noise, i.e. an agent will pick the wrong task with probability p. For $p = 50\%$ both policies are equivalent.

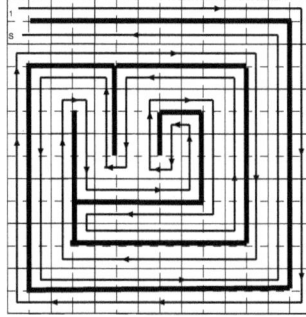

Figure 4.9: A 14x14 grid with an (arbitrary) Hamiltonian cycle constructed by the STC algorithm (Gabriely & Rimon, 2001). An optimal partitioning divides the cycle equally among all robots.

The following algorithm is defined. Initially, robots are distributed at

$$x_i(0) = i\frac{S}{N}, \qquad i = 0, \ldots, N-1 \tag{4.39}$$

where $x_i(k) = j$ denotes robot i being at vertex s_j. A robot's motion at $x_i(k) \in \mathcal{H}$ is now given by the update rule

$$x_i(k+1) = (x_i(k) + 1)\%S \tag{4.40}$$

where '%' is the modulo, and thus robot i moves along $\mathcal{H}$ in cycles. Assuming that the time for navigating and covering a vertex is τ^s, time passed after passing k nodes calculates to $k\tau^s$.

If the system is fully deterministic, the expected value for the time to completion T thus calculates to $E[T] = \frac{S}{N}\tau^s$ with variance $V[T] = 0$ as there is no redundant coverage.

In practice, however, τ^s is subject to sensor and actuator noise. Assuming τ^s to be distributed according to a truncated normal distribution with expectation $E[\tau^s]$ and variance $V[\tau^s]$, the expected time to completion T is given by

$$E[T] = E[\tau^s]\frac{S}{N}, \qquad V[T] = V[\tau^s]\left(\frac{S}{N}\right)^2 \tag{4.41}$$

using $E(aX) = aE(X)$ and $Var(aX) = a^2Var(X)$ (linear transformation).

In reality τ^s or similar cumulated noise measures are seldom normally distributed, but rather follow long-tail distributions (Gat 1995), see also Figure 6.6, page 108. In this case, and in particular if the distribution is non-parametric or if the robot controller is more complex, quantitative insight can be obtained by simulating the system by iterating (4.39) for each robot in a discrete event system simulator and sampling τ^s from its real or realistically simulated distribution.

4.4   Calibration of Model Parameters

For parameter calibration, two complementary approaches are used. Parameters (encountering probabilities and interaction times) are estimated based on the geometry of the environment and objects therein as well as a robot's sensorial configuration. As this approach is very coarse and has its limitations in particular for estimating the encountering probability in overcrowded scenarios, parameters are also estimated from systematic experiments involving either a subset of robots (one or two, e.g.) or a subset of behaviors (only collision avoidance, e.g.).

4.4.1   Parametric Calibration

Following the taxonomy from (Correll & Martinoli 2004b), we use an object's *geometric detection probability*, which is given by the ratio of its *detection area* and the total area of the arena, to estimate the *encountering probability*. The detection area A_i of object i is the area in which the object can be detected by a robot. For example, the detection area of objects from the case study considered in this thesis can be estimated from data as given in Figure 5.10, page 78. The data has been collected using a series of experiments (in the order of thousands of collisions, depending on the size of the object) in the realistic simulator *Webots* (Michel 2004). In each experiment, an Alice robot was programmed with the same collision avoidance routine that was used in the inspection experiments.

Given the detection area of object i, the geometric detection probability g_i calculates to

$$g_i = \frac{A_i}{A_{total}} \tag{4.42}$$

In order for calculating the rate of encountering an object from its geometric detection probability, the robot's mean speed v, the detection area of the smallest object A_s, and the average robot's detection width w_d is used as developed in (Martinoli et al. 2004)

$$p_i = r_i T \approx \frac{v w_d}{A_s} g_i T \tag{4.43}$$

Here, A_s is the area of the smallest object in the arena, and w_d is the robot's detection width. Thus $v w_d$ corresponds to the area that a robot is covering with its sensors per second. The robot's detection width is a function of the robot's sensorial configuration, and represents twice the average distance between robot's and object's center where detection takes place. Although one could estimate w_d given the sensor range and geometry of the objects, it can be also calibrated by an experiment like the one depicted in Figure 5.10.

In (4.43), the encountering probability is independent of the object's shape, which we validated experimentally in (Correll & Martinoli 2004b) for simple geometric shapes (rectangles and circles). In scenarios endowed with multiple obstacles, however, spatial uniformity can not always be assured. If the obstacles impose a preferential direction on the robots as was observed for the inspection experiments presented in (Correll & Martinoli 2004a), the time and location independent encountering probability becomes spatial and time-dependent.

4.4.2   Optimization of Model Parameters

Given an observation of some states of the system $N(k)$, and a model $\hat{N}(k, \hat{\Theta})$ with parameters $\hat{\Theta}$, identification of the system parameters Θ^* can be formulated as the following optimization problem (Correll & Martinoli 2006a):

$$\Theta^* = \arg\min_{\hat{\Theta}} \sum_{k=0}^{K} \left(N(k) - \hat{N}(k, \hat{\Theta}) \right)^2 \qquad (4.44)$$

with K the duration of the longest experiment. That is, we are interested in the set of parameters $\hat{\Theta}$ that minimizes the error between model prediction and observation. Equation (4.44) can be solved numerically using experimental data, which are not necessarily gathered from the full system (e.g., only a few robots), and parameters calibrated using the method from Section 4.4.1 can provide a good initial guess $\hat{\theta}_0$.

Analytical Optimization

Consider the following, linear model. A swarm of robots is moving on a bounded arena, performing collision avoidance with other robots and the boundary walls using a reactive controller, and a model for the average number of robots in the search and collision avoidance states shall be derived from experimental data. The system is modeled in discrete time, as the observed data for system identification is collected by sampling.

An Identification Experiment: An experiment is characterized over a time interval $0 \leq k \leq n$ by its state vector $N(k)$ and parameters that are set by the experimenter (e.g., the number of robots N_0). The state variables are measurements of an arbitrary metric of interest, for instance, the average number of robots searching at time k, $N_s(k)$.

A Candidate Model: Following the methodology outlined in Section 4.2, the system is modeled by a PFSM with three states: search, avoidance of walls, and avoidance of robots. This approach involves the following assumptions. Every time step, any of N_0 robots can encounter another robot with probability p_r (and any other robot with probability $p_R = p_r(N_0 - 1)$), and a wall with probability p_w. Also, a collision can be sufficiently characterized by its mean duration (T_r and T_w). This leads to the following set of difference equations:

$$\underbrace{\begin{pmatrix} N_{ar}(k+1) \\ N_{aw}(k+1) \\ N_s(k+1) \end{pmatrix}}_{N(k+1)} = \underbrace{\begin{pmatrix} 1 - \frac{1}{T_r} & 0 & p_R \\ 0 & 1 - \frac{1}{T_w} & p_w \\ \frac{1}{T_r} & \frac{1}{T_w} & 1 - p_r - p_w \end{pmatrix}}_{\theta} \underbrace{\begin{pmatrix} N_{ar}(k) \\ N_{aw}(k) \\ N_s(k) \end{pmatrix}}_{N(k)} \qquad (4.45)$$

and the initial conditions

$$\begin{pmatrix} N_{ar}(0) & N_{aw}(0) & N_s(0) \end{pmatrix}^T = (0 \quad 0 \quad N_0)^T \qquad (4.46)$$

with $N_s(k)$ being the number of robots searching at time k, $N_r(k)$ the number of robots avoiding a robot, $N_w(k)$ the number of robots avoiding a wall, and N_0 the total number of robots. The first row of (4.45) can be interpreted as follows. The number of robots avoiding another robot is decreased by those that return from a collision ($\frac{1}{T_r}N_r(k)$), and increased by searching robots colliding with another robot ($p_R N_s(k)$). The other rows of (4.45) can be interpreted in a similar fashion. Equation (4.45) can be reformulated as

$$\hat{N}(k+1)^T = N(k)^T \theta^T, \qquad (4.47)$$

where $\hat{N}(k+1)$ is the *estimate* based on the measurements of the real system $N(k)$ and the parameters θ.

Analytical Optimization: Provided the state vector measurements

$$N(k) = (N_r(k) \quad N_w(k) \quad N_s(k))^T \qquad (4.48)$$

in the interval $0 \le k \le n$, the prediction error of the model estimate $\hat{N}(k)$ can be estimated. Optimal parameters (θ) can then be found using the least-squares method

$$\min_\theta \frac{1}{n} \sum_{k=1}^{n} (N(k)^T - \hat{N}(k)^T)^2 = \min_\theta \frac{1}{n} \sum_{k=1}^{n} (N(k)^T - N(k-1)^T \theta^T)^2 \qquad (4.49)$$

With $\hat{\theta}_n$ the vector that minimizes (4.49)

$$\hat{\theta}_n = \arg\min_\theta \frac{1}{n} \sum_{k=1}^{n} (N(k)^T - N(k-1)^T \theta^T)^2 \qquad (4.50)$$

As (4.49) is quadratic in θ, the minimum value of (4.49) can be found by setting its derivative to zero:

$$0 = \frac{2}{n} \sum_{k=1}^{n} N(k-1)(N(k)^T - N(k-1)^T \hat{\theta}_n^T), \qquad (4.51)$$

yielding

$$\hat{\theta}_n = \left[\sum_{k=1}^{n} N(k-1)N(k-1)^T \right]^{-1} \sum_{k=1}^{n} N(k-1)N(k)^T, \qquad (4.52)$$

which is straightforward to compute given the availability of the measured state variables $N(k)$.

Initial Parameter Estimation: In the above experiment, measurements for all state variables (N_r, N_w, and N_s) are available. Imagine now that it is not possible to measure N_r and N_w independently from each other (this is reasonable for collisions with robots close to the wall for instance). Then, p_R and p_w cannot be estimated (4.45), but only the sum $1 - p_R - p_w$. As a work-around, additional experimental data need to be gathered by varying other parameters, for instance changing the number of robots. Such a procedure leads to an identification problem with a smaller number of degrees of freedom, but it might not be feasible to conduct it for every single parameter; in particular for systems where the ratio of parameters to the number of observed state variables is high. Then, an initial estimate using the calibration heuristic from Section 4.4 is extremely helpful.

Numerical Optimization

The system dynamics of the system under study in the previous section is linear in θ (4.51), which allows to formulate an analytic solution for θ (4.52). In a system where this is not the case, analytical solutions to (4.44) are most likely unfeasible. Numerical solutions then can be obtained by a suitable optimization algorithm (*fmincon* from the Matlab^{TM} optimization toolbox, e.g.).

In order to improve the quality of the numerical solution, the admissible parameter space (probabilities within 0 and 1, e.g.) should be appropriately reduced. Here, values obtained by parametric calibration (Section 4.4.1) provide a good initial estimate.

4.5   Discussion

The basic systems presented in this chapter are recurrent for modeling the various algorithms for distributed coverage in this thesis. The sub-chains for modeling reactive swarm-robotic systems will be combined in order to model a reactive algorithm for distributed boundary coverage. The circumstances under which the modeling assumptions of spatial uniformity and linear super-position will hold is then illustrated experimentally. The basic deliberative decision problems are extended to a suite of algorithms where robots calculate the next task, i.e. cell to cover, by various non-collaborative and collaborative deliberative policies. Here, choosing another task than that which was planned is the result of inaccurate navigation, whereas the ability to identify tasks still to do, i.e. uncovered cells in a coverage problem, has its reason in imperfect localization. Also, the potential delay of execution of the reactive behaviors, as considered in Section 4.3.2, will be illustrated on the real robot case study. Quantitative insight into the performance of the algorithms used is then obtained by using a combination of probabilistic and deterministic models and careful calibration of model parameters on the real-robotic platform.

Using optimization procedures to complement parameter calibration allows for a good match between prediction and experimental data (see for instance Correll & Martinoli 2006a) in cases where the system is at the limit of the modeling assumptions, e.g., in over-crowded scenarios or for non-uniform spatial distributions. The results obtained from such an approach have to be treated with care as the optimization procedure could well over-fit the data with potentially too simple models. In (Correll & Martinoli 2006a) optimization averages out more complex effects, e.g., a drift phenomenon observed in (Correll & Martinoli 2004a), which lead to a non-uniform distribution of the robots in the environment. In order to reach differently this level of accuracy, the level of detail in a model would need to be increased at the cost of analytical tractability, e.g., by explicitly capturing the non-uniform spatial distribution in the coverage example (see also Prorok 2006).

4.6   Conclusion

This chapter has shown how probabilistic models can be systematically developed from a deterministic robot controller's description. Abstracting events in the environment to constant encountering probabilities can lead under the assumption of spatial uniform distribution and fully reactive robot controllers to extremely compact representations of the system by means of difference equations. Instead for deliberative systems, all possible system states have to be enumerated and their probabilities calculated.

When models become analytically intractable, which is the particular case for deliberative systems with memory and an explosion of the state space, simulation is an appropriate tool. Here, the basic idea is to average over a statistically significant amount of randomly generated trajectories through the state space. Obtaining analytical solutions also becomes difficult when the interest is not only in the average performance, but also in its distribution, or when the probabilities and state durations of the system are not constant. Also in this case, average values and their distribution can be obtained by an appropriate number of simulations, where event probabilities and their durations are drawn from their distributions.

Event probabilities and event durations can be measured using parts of the real system, e.g., by measuring sensor noise on an individual robot. Alternatively, it is possible to estimate parameters, and potentially their distributions, by solving an optimization problem defined by the mismatch between model prediction and observation for a particular set of parameters. It was exemplarily shown that such an optimization problem is likely to be under-determined if the number of degrees of freedom in parameter space are not varied separately by appropriate experiments.

Chapter Summary

- Multi-robot systems with fully reactive controllers and non-spatial performance metrics can be modeled with macroscopic difference equations that keep track of the number of robots in a certain state.

- Multi-robot systems with deliberative controllers subject to noise can be modeled by explicitly enumerating the state space.

- The memory in the robot controllers (reactive or deliberative) may lead to an untractable large state space. In this case, valuable insight into the system dynamics can be obtained by simulating the system using discrete-event simulation, possibly by sampling from the probability density function of noise present in the real system.

CHAPTER 5

Reactive Algorithms for Distributed Boundary Coverage

This chapter investigates various heuristics for distributed boundary coverage, which pose only minimal requirements on the robotic platform. The size of the environment is unknown to the robots, coordination algorithms use only minimal computation, and robots communicate locally at low bandwidth.

Starting from a behavior-based controller that moves essentially randomly from blade to blade and circumnavigates a blade until a time-out expires (Section 5.1), local communication is used at selective time intervals to temporarily "mark" the blades after inspection (Section 5.2). Section 5.3 presents a controller that uses local communication during all phases of the inspection process in order to increase dispersion in the environment. The controllers are validated using systematic experiments with a swarm of up to 30 real Alice robots (Section 3.1.1) that do not have any localization abilities and provide only low-bandwidth, local communication.

The collaborative algorithms described in this Chapter can be described by the paradigm "self-organization". As defined by Bonabeau et al. (1999) for natural systems, self-organization emerges from the interplay of four ingredients: Positive feedback (e.g., amplification of behavior due to clues in the environment or provided by other individuals) and negative feedback (e.g., saturation, resource exhaustion), randomness, and multiple interactions among individuals. Although self-organization might achieve less efficient coordination than other distributed control schemes, it can provide extremely high levels of robustness on miniature robotic platforms such as those considered in this chapter. This is particularly important when moving to environments where size is a hard constraint such as inside micro machinery, which is the motivating case study in this dissertation.

Due to the complexity arising from multiple interactions and a potentially high amount of noise, it is difficult to predict the collective behavior (and performance) of a miniature robotic swarm. This is, however, necessary to make coordination approaches, based on more or less random interactions, a viable alternative for engineering such a system. This problem can be tackled by modeling the system at multiple abstraction levels, ranging from microscopic, realistic simulation to probabilistic rate equations that are derived from the individual robot behavior (Chapter 4). We illustrate this approach by incrementally developing heuristics and models for distributed coverage. Probabilistic models are then used as design tool and help to select promising control strategies.

As the environment considered in this case study consists of regularly spaced objects, algorithms developed in this chapter are also applicable to the coverage of environments with grid-based cellular decompositions.

5.1 Reactive Coverage without Collaboration

This section describes a reactive controller that *eventually* covers the whole boundary of every blade in the environment, without making use of explicit collaboration. The performance of this controller serves as a baseline for further improvements. The metric of interest shared among all abstraction levels is the time to completion, that is the time until the boundaries of all blades are covered.

5.1.1 Individual Robot Behavior

The individual robot controllers are governed by rules that switch between reactive schemes for collision avoidance, determining an objects type, and wall following. The FSM of the robot controller is depicted in Figure 5.2 and Figure 5.1 schematically depicts the behaviors.

Robots are searching through a bounded arena at random ①, and inspect objects (turbine blades) that they encounter in the environment by circumnavigating them ②. In order to identify an object's type, a robot tries to approach its contour until a specific distance is reached. The sensor reading that corresponds to this distance, however, is unfeasible to obtain for black surfaces such as the arena boundary. If the robot detects that it is unable to come close enough to an object within a certain time (0.5s in our

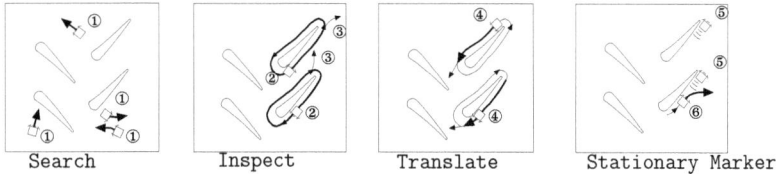

Figure 5.1: Schematic representation of the behaviors for reactive coverage without collaboration (Section 5.1) and reactive coverage using stationary markers (Section 5.2).

experiments), it abandons the object, and otherwise follows its contour clockwise or counter-clockwise (depending on the angle the object was approached with). The robots' controllers are endowed with a timeout that indicates when a blade should be left. After the timeout has expired, the robot leaves the blade when encountering its tip ③. By using a sufficiently long timeout, a robot circumnavigates the blade at least once. The rationale behind leaving a blade only at its tip stems from the fact that the tip can be easily distinguished from the rest of the blade using on-board sensors.

Using a controller that leaves blades only at their tips, will lead to a strong directional bias. For an environment as considered here, robots will soon cluster in the bottom right corner of the environment (Correll & Martinoli 2004a). This phenomenon leads to sub-optimal performance of the inspection as some blades will be inspected more often than others[1]. For this reason, the following additional behavior is implemented: robots will leave a blade only with a probability of p_t when at the tip ($p_t = 50\%$ in our experiments), and translate along the blade's contour in order to leave a blade at its round end otherwise ④.

For purely reactive behavior, robots stick to an action scheme as long as the appropriate stimulus is provided. For this reason, whenever a robot abandons inspection, avoids an obstacle, or another robot, collision avoidance is enforced for 2s.

5.1.2 Microscopic Models

The robot behavior described in Section 5.1.1 has been implemented in Webots. Webots faithfully models not only the behavior but also individual sensors and actuators of each robot. On a higher abstraction level, the individual robot can be described by the PFSM depicted in Figure 5.3 (see Section 4.2, page 40 for an introduction to the concept). A robot can be searching (N_s), avoiding a robot (N_r) or a wall (N_w), inspect a virgin

---

[1]In a real turbine, boundaries would be non-existent, and the combination of environmental template and robot behavior could lead to improved exploration of the environment

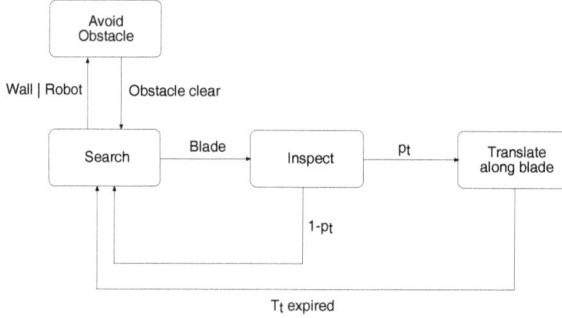

Figure 5.2: FSM of a reactive controller for randomized coverage. A robot is randomly walking and avoids obstacles. Upon encountering a blade, the robot follows its contour until a time-out expires.

(N_v) blade, inspect a partly inspected blade (N_p), i.e. inspecting a blade concurrently with another robot that considers this blade a virgin blade, inspect an already inspected blade (N_i), or translate back along a blade after inspection (N_t). The three inspection states are considered one state in the FSM of the individual robot (Figure 5.2) and are required for the allocentric metric "time to completion". The probability that a robot encounters either a virgin, partly inspected, or inspected elements, is then a function of their numbers. Distinguishing between virgin and partly inspected blades will become important when considering mobile marker-based collaboration in Section 5.3.

The state transition probabilities are given by the encountering probabilities for a blade p_e, a wall p_w, another robot p_r, and the probability to translate along a blade after inspection p_t. The state durations are given by T_e for inspecting a blade, T_w for robot-wall collision, T_r for robot-robot collision and T_t for translating along a blade. The encountering probabilities p_e, p_w, and p_r as well as the interaction times T_e, T_w and T_r are model parameters that reflect physical properties of the system, whereas p_t and T_t are control parameters of the individual robot. Parameters are summarized in Table 5.1.

N_0 instances of such an automaton (one for each robot) can now be simulated using synchronous agent-based simulation; the behavior of the whole can then be observed with respect to a certain metric, here the number of virgin blades at time k. Initially, all M_0 blades are virgin. As soon as a robot starts inspecting a virgin blade it is considered "under inspection", and "inspected" after time T_e. If two or more robots inspect the same blade, only the first robot inspects a virgin blade and all others inspect a partly inspected blade. For simplicity, robots maintain their status until they finish inspection, even if the blade itself is considered fully inspected after the first robot leaves.

System Parameters	Description	Section
p_e, T_e	Encountering probability of a blade and average time for inspection.	5.1.2
p_r, T_r	Encountering probability for a robot and average time for resolving a robot-robot collision.	5.1.2
p_w, T_w	Encountering probability for a wall and average time for resolving a robot-wall collision.	5.1.2
N_0, M_0	Total number of robots and total number of blades, respectively.	5.1.2
γ	Coupling parameter for collaboration, $\gamma = 0$ corresponds to no communication, $\gamma = 1$ corresponds to communication over the whole detection area of a robot	5.3.2
Control Parameters		
p_t, T_t	Probability to translate along a blade after inspection, and time required to do so.	5.1.2
$T_{m,init}$, k_m, Δk_m, $\tilde{T}_m(k)$	Parameters defining the profile of $T_m(k)$, the time a robot acts as stationary marker	5.2.4.

Table 5.1: Summary of model parameters used in this Chapter

5.1.3 Macroscopic Model

$N_s(k)$ represents the number of searching robots. As the number of robots N_0 is constant, the number of searching robots calculates to

$$N_s(k) = N_0 - N_r(k) - N_w(k) \tag{5.1}$$
$$-N_v(k) - N_p(k) - N_i(k) - N_t(k)$$

where $N_r(k)$, $N_w(k)$, $N_v(k)$, $N_p(k)$, $N_i(k)$, and $N_t(k)$ are the number of robots in avoidance (robots and walls), inspecting a virgin, a partly inspected or a fully inspected blade, and translating along a blade after inspection, respectively.

$N_r(k)$, and $N_w(k)$ are delay states and can hence be calculated according to (4.6) in Section 4.2.1.

$$N_r(k+1) = N_r(k) + p_R N_s(k) - \frac{1}{T_r} N_r(k) \tag{5.2}$$

$$N_w(k+1) = N_w(k) + p_w N_s(k) - \frac{1}{T_w} N_w(k) \tag{5.3}$$

Here, $p_R = (N_0 - 1)p_r$ and p_w are the encountering probabilities for any other robot and the wall, respectively. T_r and T_w are the interaction times for robot-robot collision and collisions with the arena boundaries, respectively. For instance, the number of robots avoiding a robot are increased by searching robots encountering any other robot and decreased by those that have spent an average time T_r in N_r.

The number of robots inspecting a blade can be calculated similarly as the number of robots avoiding a collision, whereas the number of potential objects to interact with (encountering probability p_e) is not constant but time-varying:

$$N_v(k+1) = N_v(k) + p_e M_v(k) N_s(k) - \frac{1}{T_e} N_v(k) \tag{5.4}$$

$$N_p(k+1) = N_p(k) + p_e M_p(k) N_s(k) - \frac{1}{T_e} N_p(k) \tag{5.5}$$

$$N_i(k+1) = N_i(k) + p_e M_i(k) N_s(k) - \frac{1}{T_e} N_i(k) \tag{5.6}$$

$M_i(k)$ denotes the number of inspected blades, which are given by $M_i(k) + M_p(k) + M_v(k) = M_0$ (the number of blades is constant), and $M_p(k)$ and $M_v(k)$ at time k are given by

$$M_v(k+1) = M_v(k) - p_e M_v(k) N_s(k), \quad M_v(0) = M_0 \tag{5.7}$$

$$M_p(k+1) = M_p(k) + p_e M_v(k) N_s(k) - \frac{1}{T_e} N_v(k) \tag{5.8}$$

For instance, the number of virgin blades is reduced by the number of searching robots encountering a virgin blade (with probability $p_e M_v(k)$).

The number of robots translating along a blade after inspection $N_t(k)$ calculates slightly differently. While the state duration for inspection is variable and is expressed by its average T_e, as it depends on the location of where the robot attached to a blade, the time for translating back along a blade after inspection is a function of the length between a blade's tip and its round end. This information is then used to hand-code a time-out (time T_t) in the controller. Thus, the transient dynamics of the system are more faithfully represented by opting for a fixed delay:

$$N_t(k+1) = N_t(k) + p_t \frac{1}{T_e} \left(N_v(k) + N_p(k) + N_i(k) \right) \tag{5.9}$$
$$-p_t \frac{1}{T_e} \left(N_v(k-T_t) + N_p(k-T_t) + N_i(k-T_t) \right)$$

The number of robots in N_t is increased by all robots leaving an inspection state after time T_e, and decreased by the number of robots that have entered N_t before T_b time steps. The factor p_t reflects the fact that only some of the robots need to translate back along the blade after inspection in order to maintain an uniform distribution. As the tips of the blades all point in the same direction in this case study, p_t is set to 0.5.

Inspection is completed if all blades are inspected ($M_i(n) = M_0 - \epsilon$), with $0 < \epsilon$ a certain degree of confidence. To compute the time to completion nT, $M_i(n) = M_0$ is an easy condition to apply in the experiment. However, in the macroscopic model, this represents a limit condition as $\lim_{k \to \infty} M_i(k) = M_0$, and thus we solve the macroscopic model for $M_i(n) = M_0 - \epsilon$, with ϵ a reasonable small value. Notice, that $M_i(k)$ does *not* represent the fraction of completed experiments at time interval k, which would allow for calculating the expected value $E[n] = \sum_{k=0}^{\infty} \dot{M}_i(k)/M_0 k$. Instead, $M_i(k)$ represents the fraction of blades covered at time k.

Initial Conditions

The initial conditions are $N_s(0) = N_0$ and $N_r(0) = N_w(0) = N_v(0) = N_i(0) = N_t(0) = 0$ for the robotic system (all robots in search mode) while those of the environmental system are $M_v(0) = M_0$ and $M_p(0) = M_i(0) = 0$ (all blades virgin). As customary for time-delayed DE, we assume $N_x(k) = M_x(k) = 0$ for $k < 0$.

Analysis of Convergence

Convergence is shown by proving the existence of a stationary distribution for the robotic system. The stationary distribution of the robotic system then allows for obtaining an explicit solution for the environmental system, i.e. the number of covered blades.

Theorem 5.1.1. *The Markov chain describing the state distribution (Section 4.1.7) of the robotic system depicted in Figure 5.3a will reach a stationary distribution for $\lim_{k \to \infty}$.*

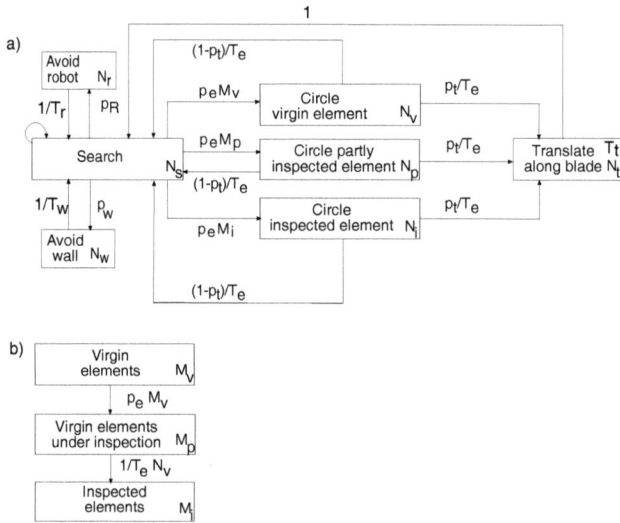

Figure 5.3: a) Graphical representation of the rate equations that are derived from the Markov chain modeling the individual robot's state, b) rate equations modeling the environmental state (the number of inspected elements). State transitions of the environmental model are a function of the state variables of the robotic system.

Proof. By inspection, the Markov chain is irreducible, aperiodic, and does not contain transient states and is thus recurrent. As the average return time for each states is finite ($T_t < \infty$ can be safely assumed), the state space of Figure 5.3a is finite. Using Theorem 4.1.1, the Markov chain thus reaches a stationary distribution. $\quad\square$

Given a stationary distribution for the number of searching robots $N_s^* > 0$ using Theorem 5.1.1, (5.7) can be solved (see Section 4.2.2) to

$$M_v(k) = M_0(1 - p_e N_s^*)^k \qquad (5.10)$$

which converges asymptotically to zero for $k \to \infty$ as $0 < p_e N_s^* < 1$ by definition (linear super-position of encountering probabilities, Section 4.1.5).

5.2 Reactive Coverage with Stationary Marker-Based Collaboration

Intuitively, coverage performance might be increased by "marking" covered areas (see also the work of Svennebring & Koenig (2004) and references therein). As the Alice platform has no possibility to do so, using the robot itself as a marker is an option. Acting as a marker is then a trade-off between contributing to task progress by coverage of potentially uncovered areas and preventing other robots from performing redundant coverage.

Correll & Martinoli (2005) show for this case study that waiting for a constant amount of time, *never* speeds-up coverage and it is always beneficial to continue coverage rather than to collaborate. This is not the case, when using a *dynamic* marker policy, that is changing the policy over time (Correll & Martinoli 2006b).

The remainder of this section will first describe the individual robot behavior for marker-based collaboration (Section 5.2.1), and then extend the probabilistic models developed for the basic robotic controller (Section 5.1.2 and 5.1.3) to model collaboration using the generic collaboration model from (Martinoli et al. 2004), see also Section 4.2.3, page 45. Finally, an optimal marker policy is found using optimal control techniques.

5.2.1 Individual Robot Behavior

The robot behavior for marker-based collaboration is summarized by the FSM depicted in Figure 5.4 and the schematic representation of Figure 5.1. As in the non-collaborative policy (Section 5.1), the robot searches for blades throughout the turbine, combining schemes that drive the robot forward, avoid obstacles, and follow contours. For temporarily marking blades, the following behavior is added: after inspecting a blade, a

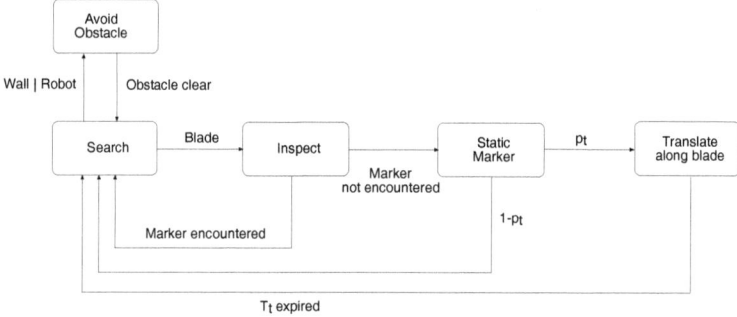

Figure 5.4: FSM of a robot controller for randomized coverage with marker-based collaboration. After inspecting a blade found by random walk, a robot rests as its tip for a certain time and prevents following robots from finishing inspection.

robot marks a blade by resting at its tip for the time $T_m(k)$ ⑤ and signals to all robots within its range possibly approaching this blade to avoid it, or abort its circumnavigation, in case the other robot has already attached to the blade at another point ⑥. The time-out $T_m(k)$ follows a profile that is hard-coded into the robot's controller.

5.2.2   Microscopic Model

The Probabilistic Finite State Machine for the marker-based collaboration behavior is depicted in Figure 5.5. An additional state compared with Figure 5.3 has been introduced for robots marking blades after inspection. Robots remain in this state for time $T_m(k)$. Upon encountering a resting robot, an inspecting robot immediately leaves the current blade. The probability for inspecting a blade that is marked by a robot is given by the ratio of markers and inspected blades $N_m(k)/M_i(k)$ (see below for the definition of $N_m(k)$). If a robot inspects a marked blade, it will encounter the marker after T_f time steps on average. The probability for leaving an already inspected blade is thus given by $\frac{N_m(k)}{T_f M_i(k)}$.

The time $T_m(k)$ spent for marking a blade is given by

$$T_m(k+1) = T_m(k) + \tilde{T}_m(k), \tag{5.11}$$

with $T_m(0) = T_{m,init}$, and $\tilde{T}_m(k) = 0$ if the marker policy does not change, while $\tilde{T}_m(k) = f(k)$ with $f(k)$ a non-linear function with arbitrary parametrization in the dynamic case, allowing to increase or decrease the time spent as a marker during the experiment.

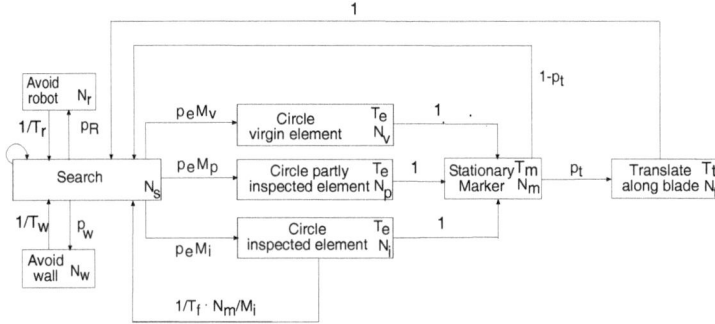

Figure 5.5: Graphical representation of the difference equation system for marker-based collaboration. Notice the additional state `Marker` and the additional transition from `Circle inspected element` back to search.

5.2.3   Macroscopic Model

The dynamics of the PFSM shown in Figure 5.5 at system level are given by the following difference equations. For M_0 blades and N_0 robots, the number of robots in obstacle avoidance N_r and N_w are given by (5.2) and (5.3). Also, equations modeling the inspection progress M_v, M_p and M_i are given by (5.7)–(5.8) with $M_i(k) = M_0 - M_v(k) - M_p(k)$. Instead, the number of robots covering virgin, partly and fully inspected blades N_v, N_p and N_i, the number of robots marking a blade N_m, the number of robots translating back along the blade's contour N_t, and the number of robots in search mode N_s are modified due to the marker policy introduced in this section.

The number of robots covering an already inspected blade calculates to

$$N_i(k+1) = \qquad N_i(k) + p_e M_i(k) N_s(k) - \frac{N_m(k)}{T_f M_i(k)} N_i(k) \qquad (5.12)$$

$$-p_e M_i(k - T_e) N_s(k - T_e) \Gamma(k - T_e; k)$$

As opposed to (5.6), the inspection state can now additionally be quit when encountering a marker. The probability for inspecting a marked blade given that the blade is already inspected is $\frac{N_m(k)}{M_i(k)}$, and the average time until a marker is reached is T_f. The Γ-function (see also Section 4.2.3, page 45) represents the fraction of robots that unavailingly waited for collaboration. In the "stick-pulling experiment" (Martinoli et al. 2004), Γ represents the fraction of robots that did not find another robot to help with their task, pulling sticks out of the ground. Here instead, Γ expresses the fraction of robots that did not encounter a beacon before leaving a blade after T_e.

$$\Gamma(k - T_e; k) = \prod_{j=k-T_e}^{k} (1 - \frac{N_m(j)}{M_i(j)T_f}) \tag{5.13}$$

As $\Gamma(k - T_e; k)$ is defined for constant T_e, the number of robots inspecting virgin blades $N_v(k)$ and partly inspected blades are now modeled accordingly as delay states with constant duration T_e, and write

$$N_v(k+1) = N_v(k) + p_e M_v(k)N_s(k) - p_e M_v(k - T_e)N_s(k - T_e) \tag{5.14}$$

$$N_p(k+1) = N_p(k) + p_e M_p(k)N_s(k) - p_e M_p(k - T_e)N_s(k - T_e) \tag{5.15}$$

The number of robots marking a blade is given by

$$
\begin{aligned}
N_m(k+1) = {} & N_m(k) + p_e M_i(k - T_e)N_s(k - T_e)\Gamma(k - T_e; k) \\
& + p_e M_v(k - T_e)N_s(k - T_e) \\
& - \sum_{j=0}^{-\tilde{T}_m(k)} p_e[M_v(k - T_e - T_m(k)) + M_p(k - T_e - T_m(k))]N_s(k - T_e - T_m(k) + j) \\
& - \sum_{j=0}^{-\tilde{T}_m(k)} p_e M_i(k - T_e - T_m(k))N_s(k - T_e - T_m(k) + j)\Gamma(k - T_e - T_m(k) + j; k)
\end{aligned}
\tag{5.16}
$$

The summations introduced in (5.16) are necessary as $T_m(k)$ is time variant. In short, if T_m is decreased by $\tilde{T}_m$, all robots that became a marker in the interval $[k - T_m(k) - \tilde{T}_m(k); k - \tilde{T}_m(k)]$ need to continue searching at once. For increasing T_m, no robot shall leave the marker state for $\tilde{T}_m$. Note, that using this notation, the model can only give valid prediction for $\tilde{T}_m(k)\epsilon] - \infty; 1]$ as for $\tilde{T}_m(k) > 1$, $\tilde{T}_m(k+1)$ will be zero, and thus renders the sum useless.

Finally, the number of searching robots is given by

$$
\begin{aligned}
N_s(k+1) = {} & N_0 - N_v(k+1) - N_i(k+1) - N_p(k+1) \\
& - N_r(k+1) - N_w(k+1) - N_m(k+1) - N_t(k+1)
\end{aligned}
\tag{5.17}
$$

Initial Conditions

The initial conditions are the same as those in Section 5.1.3. Additionally, the profile $T_m(k)$ is initialized by $T_m(0) = T_{m,init}$, and $N_m(0) = 0$.

Analysis of Convergence

Convergence of the algorithm can be shown using the same reasoning as in Section 5.1.3 for $T_m < \infty$. Thus, for $T_t < \infty$ and $T_m < \infty$, the stationary marker-based algorithm always converges to complete coverage.

5.2.4 Dynamic Optimization

Coverage performance can be improved by finding an optimal marker-activation policy, i.e. a profile for $T_m(k)$ that minimizes time to completion nT. Then $\tilde{T}_m(k)$ is the time-varying decision variable that defines the profile $T_m(k)$. The optimization problem is formulated as follows

$$\min_{\tilde{T}_m(0)...\tilde{T}_m(N-1)} J = nT \tag{5.18}$$

$$s.t. \qquad 0 = M_i(n) + \epsilon - M_0$$

where $0 = M_i(n) - M_0 + \epsilon$ is a terminal constraint ensuring all blades have been inspected and ϵ is a very small value.

For simplifying the dynamic optimization problem $\tilde{T}_m(k)$ is parameterized as follows:

$$\tilde{T}_m(k) = \begin{cases} a & \text{when } k_m \leq k \leq k_m + \Delta k_m, \\ 0 & \text{else.} \end{cases} \tag{5.19}$$

The profile $T_m(k)$ is hence (see equation 5.11) defined by three decision variables $T_{m,init}$, k_m and Δk_m, and a fixed parameter a (see below). Thus, $T_m(k)$ has the value $T_{m,init}$ until time k_m where it increases until $T_m(k) = T_{m,init} + a\Delta k_m$ at $k_m + \Delta k_m$. The optimization problem can hence be reformulated to

$$\min_{T_{m,init}, k_m, \Delta k_m} J = nT \tag{5.20}$$

$$s.t. \qquad 0 = M_i(n) - M_0 + \epsilon$$

with J the cost and $u = [T_{m,init}, k_m, \Delta k_m]$ the vector containing the decision variables to optimize.

5.3 Reactive Coverage with Mobile Marker-based Collaboration

The basic coordination approach (Section 5.1.1) benefits from parallel task execution, whereas randomness and coupling among the robots are limited to physical interactions between robots and the environment, and robots that repel each other physically, respectively. In Section 5.2.1 coupling between robots has been introduced by robots acting as markers. In this section, robots combine marking with translating behavior, leading to a mobile marker-based policy.

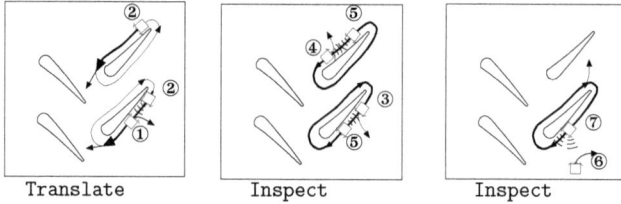

Figure 5.6: Local communication during the `Translate` and `Inspect` states yield improved dispersion of the robots and reduced redundancy.

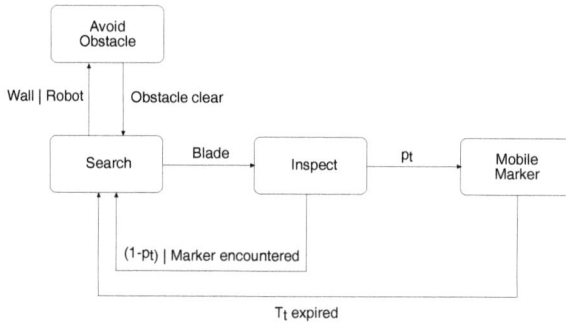

Figure 5.7: FSM of a robot controller performing reactive, mobile marker-based collaborative coverage. Robots inspect blades encountered during random walk. Local communication is used for abandoning blades currently inspected by other robots which act as mobile markers and for promoting dispersion in the environment.

5.3.1 Individual Robot behavior

In order to further reduce redundant coverage, the following additional behaviors that exploit communication among the robots are introduced (Figure 5.6): First, inspecting robots ① that encounter a robot translating back along an already inspected blade ② will abort inspection. Second, robots abandon an inspection if they follow ③ or encounter ④ another inspecting robot ⑤ (in this case only the robot having the blade to its right hand side will leave). Third, searching robots ⑥ will not attach to a blade if there is an inspecting robot ⑦ nearby (in whatever direction).

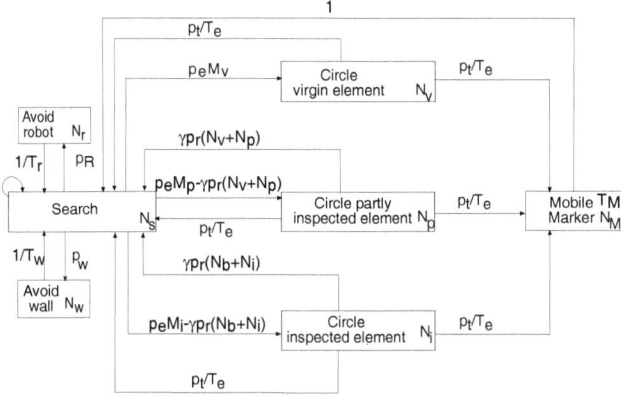

Figure 5.8: Graphical representation of the macroscopic model for mobile marker-based collaborative coverage. Notice the additional branches from `Circle inspected element` and `Circle partly inspected element` to `Search` state, when compared with the PFSM without collaboration (Figure 5.3).

5.3.2 Microscopic Model

The PFSM including communication between robots is depicted in Figure 5.8. The PFSM keeping track of the number of blades being inspected is identical to that from Figure 5.3, b.

The PFSM in Figure 5.8 introduces the following changes: The detection probabilities of blades that have already been visited by at least one robot are reduced by the encountering probabilities of other robots already inspecting the blade or translating along its contour after inspection. Also, robots can now leave a blade upon encountering another robot on its contour. As the blade shields infra-red communication in one direction, the encountering probability of a robot during inspection is reduced by a factor $\gamma = 0.5$. Consequently, for $\gamma = 0$, i.e. no communication, the PFSM from Figure 5.8 reduces to that of Figure 5.3.

5.3.3 Macroscopic Model

The collaboration among robots does not affect robot-robot (5.2) and robot-wall collisions (5.3). Also, equations that model the environmental states (5.7)–(5.8) and the number of robots in search (5.17) can be maintained. The equation system for the number of inspecting robots (5.4)–(5.6), however, needs to take into account repulsion by other robots, and additional state transitions for quitting redundant inspection:

$$N_v(k+1) = N_v(k) + p_e M_v(k) N_s(k) - \frac{1}{T_e} N_v(k) \tag{5.21}$$

$$\begin{aligned} N_p(k+1) = {} & N_p(k) + (p_e M_p(k) - \gamma p_r \left(N_v(k) + N_p(k)\right)) N_s(k) \\ & - \frac{1}{T_e} N_p(k) \\ & - \gamma p_r \left(N_v(k) + N_p(k)\right) N_p(k) \end{aligned} \tag{5.22}$$

$$\begin{aligned} N_i(k+1) = {} & N_i(k) + (p_e M_i(k) - \gamma p_r \left(N_M(k) + N_i(k)\right)) N_s(k) \\ & - \frac{1}{T_e} N_i(k) \\ & - \gamma p_r \left(N_M(k) + N_i(k)\right) N_i(k) \end{aligned} \tag{5.23}$$

For modeling repulsion from blades, in (5.22) the detection area $p_e M_p(k)$ of elements already under inspection is reduced by the detection area of robots currently covering a virgin blade $\gamma p_r(N_v(k) + N_p(k))$. Similarly, in (5.23) the detection area $p_e M_i(k)$ of already inspected elements is reduced by the detection area $\gamma p_r(N_M(k)+N_i(k))$ of robots performing redundant inspection and translating back along the blade.

Analysis of Convergence

For $\gamma > 0$, convergence can be shown using the same reasoning as used in Section 5.1.3 (for $\gamma = 0$ both systems are equivalent). As $N_p(k)|_{\gamma>0}$ and $N_i(k)|_{\gamma>0}$ will increase slower and decrease faster than $N_p(k)|_{\gamma=0}$ and $N_i(k)|_{\gamma=0}$ which can be seen by inspection of the associated difference equations, and all other states are not affected from γ, one can see that $N_s^*|_{\gamma>0} > N_s^*|_{\gamma=0}$. Thus $M_v(k)|_{\gamma>0}$ (5.10) will converge faster to zero than $M_v(k)_{\gamma=0}$.

5.4   Results

A first series of experiments validates the modeling assumptions. Model parameters are then determined by exploiting information about the geometry of the environment and the robot's speed and sensory configuration, as well as by numerical optimization. Model prediction of the macroscopic model and Webots achieves good agreement with real robot experiments. The calibrated parameters are then used to find an optimal stationary marker policy. Finally, model prediction and experimental results for non-collaborative and mobile marker-based coverage are compared. The time discretization of all models is $T = 1s$.

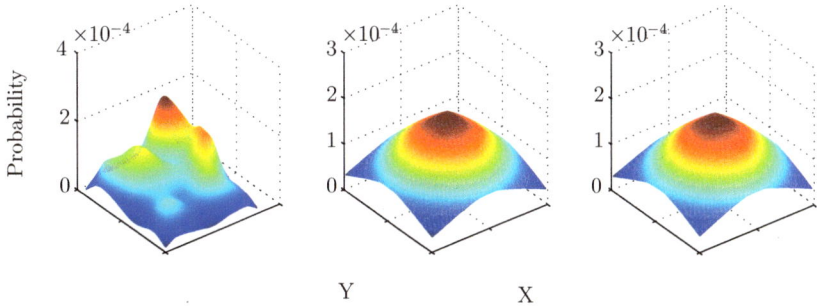

Figure 5.9: Estimated probability density function for the location within the arena for 20 robots over around 30 experiments when leaving the blade only at the tip (*left*), or translating back along the blade in 50% of the cases for non-collaborative and mobile marker-based inspection (*middle* and *right*, respectively). The two-dimensional probability density functions are discretized (100x100 grid) and low-pass filtered.

5.4.1 Validation of Modeling Assumptions

In order to validate the assumption of spatial uniform distribution of robots in the environment, the trajectories of 20 robots in 30 coverage experiments are analyzed. The position of each robot has been recorded and attributed to a two-dimensional grid discretization of the environment (100 x 100 bins). The resulting 2D-histogram has been normalized by the total number of measurements, leading to an estimate of the probability density function of a robot's location in the arena.

Figure 5.9*a* shows the spatial distribution that results from a controller where a blade is only left at its tip (Correll & Martinoli 2004*a*). Figure 5.9*b* and Figure 5.9*c* show the resulting spatial distribution if robots are leaving at either end with equal chance for a system with and without communication.

5.4.2 Parameter Calibration

Encountering probabilities for obstacles, blades and walls used in the microscopic and macroscopic model have been obtained by numerically solving the optimization problem (4.44) for 100 experiments with *non-collaborating* swarms of 10, 20, and 30 robots, and an initial estimate provided by (4.43), page 54, and data from Figure 5.10. The detection width (twice the distance from center to center of a robot upon detection) has been set to $w_d = 0.10m$ (compare Figure 5.10, bottom row, center) and $v = 0.04\frac{m}{s}$, with $A_s = A_r$.

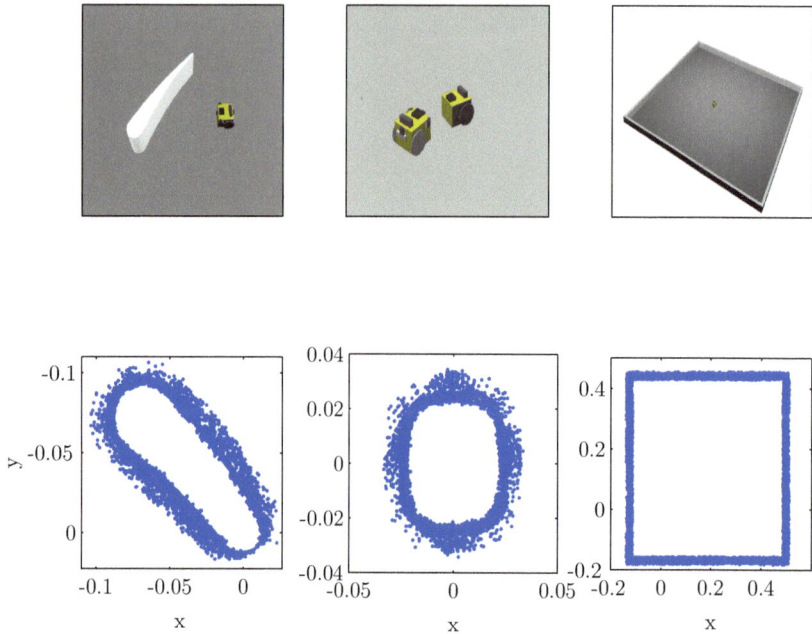

Figure 5.10: Experimental setup for determining the detection area of a blade (*left*), another robot (*middle*), and the outer walls (*right*) in a realistic simulator (*top row*). The resulting detection area given by all points where the object has been detected by a robot (*bottom row*).

The speed $v = 0.04\frac{m}{s}$ is the speed the robot reaches in obstacle free areas of the arena and thus can be considered an upper bound.

Detection areas of all objects, the initial estimates and the encountering probabilities actually used are summarized in Table 5.2. The interaction times are given in Table 5.3.

Values for the interaction times base on kinematic considerations (distance is equal to speed times time, e.g.) and observations of the real system and are summarized in Table 5.3. Interaction times have been kept constant during optimization in order to reduce the degrees of freedom of the optimization process. This is important as interaction times are tightly related to encountering probabilities: short interaction times with high encountering probabilities might lead to an identical prediction for the steady-state as long interaction times with low encountering probabilities.

Table 5.2: Encountering probabilities used for microscopic and macroscopic model. Initial estimate (middle column) are calculated based on the detection area of an object (left column), and value after optimization (right column). Divergences between initial estimate and optimal parameters reflect violations of the modeling assumptions and inaccuracies of the model.

	Area		$\hat{\theta}_0$	θ^*
A_r	0.0044m^2	p_r	0.010	$6 \cdot 10^{-4}$
A_e	0.0095m^2	p_e	0.022	$2 \cdot 10^{-3}$
A_w	0.0396m^2	p_w	0.093	$5 \cdot 10^{-2}$

Table 5.3: Interaction times used for microscopic and macroscopic model. Initial estimates base on trials with an individual robot or geometric considerations, and have been kept constant during the parameter optimization process.

T_r	4s	T_e	15s	T_w	4s
T_t	8s	T_f	8s		

Prediction of the macroscopic model using the calibrated parameters for the average number of inspected blades over time are now compared with experimental data from non-collaborating swarms of 10, 20, and 30 robots, and Webots in Figure 5.11 (100 experiments per team size both for real robots and Webots experiments).

5.4.3 Optimal Control of a Stationary Marker-Based Collaboration policy

For limiting the search space of the optimization problem, the parameter a (compare Equation 5.19) is constrained to take discrete values in the interval $[-1; +1]$, where $a = 0$ corresponds to a time-constant marker policy. Also, the optimization considers only two dynamic policies, first to relax a marker policy by decreasing T_m over time, and second, to foster a marker policy by increasing T_m. The evolution of the state variables for the latter case with $k_m = 40s$ is depicted in Figure 5.12. Notice in particularly the number of robots acting as markers N_m (*middle*).

Optimization using *fmincon* provided by the Matlab Optimization toolbox for various initial conditions and model parameters from Table 5.2 yields $T^*_{m,init} = 0$ and $\Delta k^*_m = 0$ for $a = -1$, i.e. starting with a marker policy always decreases performance. On the other hand, setting $a = +1$, yields an optimal policy $T^*_{m,init} = 0s$, $k^*_m = 55s$, $\Delta k^*_m = 1.8$ for $N_0 = 45$ robots. However, performance is improved only marginally by 0.7% for the chosen model parameters. In fact, whether a marker-based policy can lead

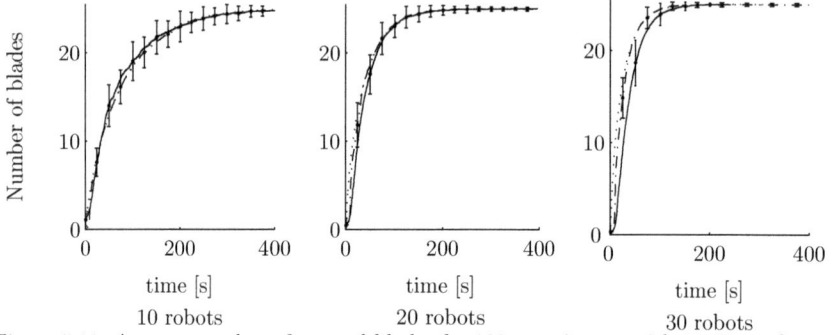

Figure 5.11: Average number of covered blades for 100 experiments with 10, 20, and 30 robots (—), realistic simulation (– –), and prediction of the macroscopic model for non-collaborative inspection ($\cdots$) after parameter optimization. Error bars depict standard deviation.

to an improvement or not appears to be a function of the time T_e effectively needed for inspection ($T_e = 15s$ in our case study). We therefore performed optimization for $10 \leq N_0 \leq 40$ robots with $M_0 = 25$ for $T_e = 75s$ (Figure 5.13). Notice that the blades in the experimental setup from (Correll & Martinoli 2004a) and (Correll & Martinoli 2006b) are much larger (factor 2.5 in length) than those used in the experimental setup of this dissertation. As the size of the robots did not change, inspection of a single blade takes also considerably longer in (Correll & Martinoli 2004a), and therefore also the time a marker-policy might save increases. This is also the case if the inspection sensor requires the robots to move slow when circumnavigating a blade.

As the performance gain for the optimal marker policies in our experimental setup is very small (around 5% compared with a system without collaboration in Correll & Martinoli 2004a), it requires large number of experiments in order to show statistically significant results (in the order of 50-100 runs), and we thus limit our experimental validation to the mobile marker-based controller.

5.4.4   Mobile Marker-Based Inspection

Results from mobile-marker based inspection for 20, 25, and 30 robots are shown in Figure 5.14. The actual improvement due to the mobile marker-based policy for real experiments (Correll, Rutishauser & Martinoli 2006), and the prediction of the macroscopic

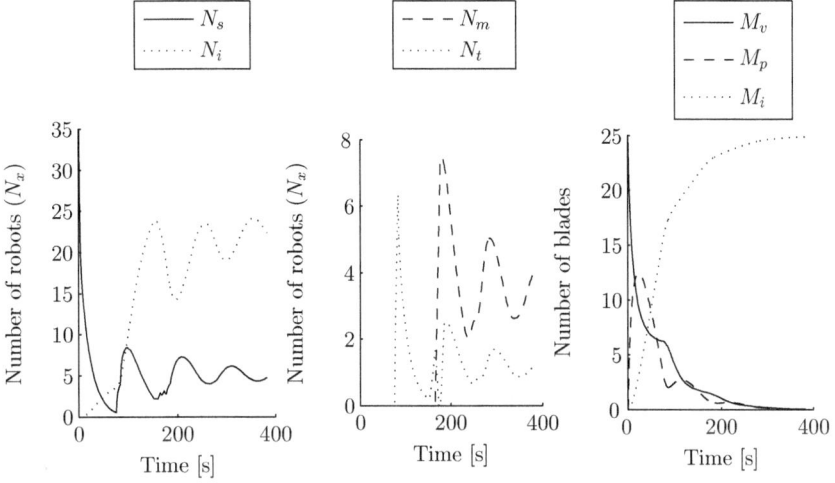

Figure 5.12: Population dynamics of the robots searching and inspecting a blade (*left*) and marking a blade and translating back along its contour (*middle*). Coverage progress is shown *right*. $T_{m,init} = 163$, $\Delta k_m^* = 12$ for $T_e = 75s$ and $N_0 = 35$.

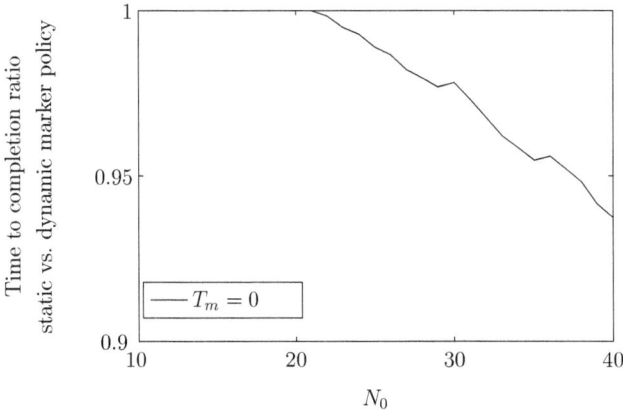

Figure 5.13: Relative improvement of a marker-based collaboration policy over non-collaborative, reactive coverage (macroscopic model) for inspection time $T_e = 75s$.

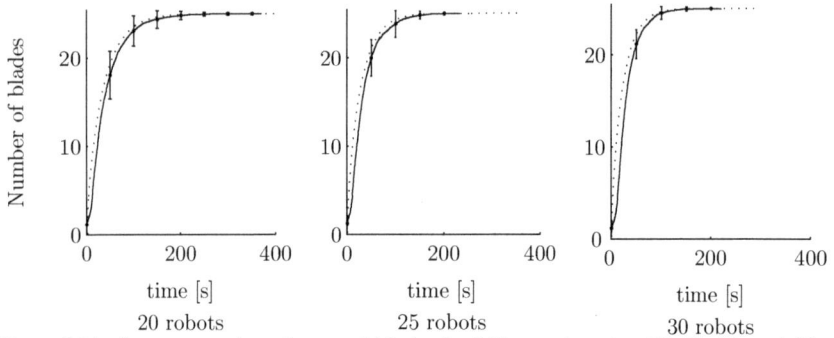

Figure 5.14: Average number of covered blades for 100 experiments with 20, 25, and 30 robots (–), and prediction of the macroscopic model for mobile marker-based inspection (···). Error bars depict standard deviation.

model[2] are shown in Figure 5.15. As the macroscopic model approaches completeness only asymptotically, the time needed to cover $M_0 = 24.9999$ blades serves as baseline for comparison. The improvement of collaboration can also be seen by looking at the performance histogram (Figure 5.16) for 100 real robot experiments (*left*) and 100 runs of synchronous, agent-based simulation of the system's PFSM (*right*).

5.5   Discussion

5.5.1   Modeling Assumptions

The inspection scenario touches the limits of the modeling assumptions (Section 4.1.4) due to (1) a high density of robots and objects, (2) the structure of the environment that influences the spatial distribution of the robots for some controllers, and (3) the geometry of the blade objects which lead to an encountering probability that is dependent on the angle of approach (a robot is unlikely to recognize a blade as such when approaching from on its ends).

The high density of robots and blades in our experimental setup clearly violates our assumption of *linear super-position* of encountering probabilities (see Section 4.1.5, page 38). In fact, this is the case as soon as detection areas of objects overlap, e.g. when a robot follows the contour of a blade. In extreme cases of over-crowding, our modeling assumption might lead to an overall encountering probability larger than one, which is

---

[2]the experimental results from mobile marker-based inspection have not been taken into account for calibration of modeling parameters

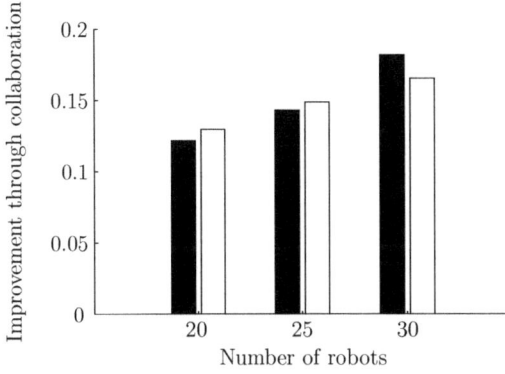

Figure 5.15: Relative improvement due to collaboration in 100 experiments with 20, 25, and 30 robots (■), and prediction of the macroscopic model (□).

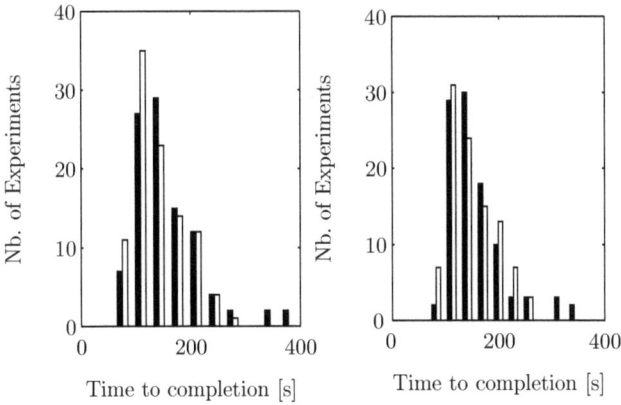

Figure 5.16: Histogram of the time to completion with (□) and without collaboration (■) for 20 robots and 100 real-robot experiments *left*, and 100 simulations of the agent-based microscopic model *right*.

physically meaning-less. Thus, in over-crowded scenarios, the encountering probability for an individual object needs to be adopted to a value usually lower than that given by its detection area, which is well reflected by our results from Table 5.2.

After implementing the additional behavior for translating along the blades, results show that the assumption of *spatial uniformity* is indeed reasonable (Figure 5.9). Nevertheless, robots show a slight bias for being in the center of the arena. Such a bias will lead to a lower probability of inspection for blades in the boundary of the arena. Using the translating behavior for 50% of the inspected blades diminishes this effect considerably.

Although robots in this experimental setup are uniformly distributed on the long run, this is not the case in many potential swarm-robotic applications that involve environmental gradients, such as in multi-robot search or when operating under the influence of wind or current. In this case, the encountering probability of an object is a function of the location in the environment, and it is necessary to not only model the average number of robots in a certain state, but also their spatial distribution (see also (Prorok 2006) for an attempt in this direction.

5.5.2   Parameter Calibration

Due to reasons mentioned in Section 5.5.1, quantitative agreement between model prediction and experimental data can only be achieved by parameter estimation (Section 5.4.2). Without this step, which has not been necessary in the seed aggregation (Martinoli et al. 1999, Agassounon et al. 2004) and stick-pulling (Martinoli et al. 2004) experiments, models still allow for good qualitative predictions, which enables proofs of desired properties in a probabilistic sense.

Using techniques from system identification for achieving good quantitative agreement between model prediction and experimental data then allows for further analysis and optimization of the system such as the design of stationary and mobile marker-based collaboration algorithms (Sections 5.4.3 and 5.4.4). Also, quantitative correct models allow for estimating the distribution of the performance (Figure 5.16), which is important for providing confidence intervals of the performance and thus enables fully reactive/randomized control architectures as a viable alternative for engineering multi-robot systems.

The predictive power gained from models that have been calibrated using only a limited set of experimental data remains an open question (see also the discussion in Section 4.5). For this case study, estimation of encountering probabilities from experiments with non-collaborating teams of 10, 20, and 30 robots have been sufficient for faithfully predicting performance gain of mobile marker-based collaboration policies (Figure 5.15).

How many parameters of the system need to be excited (in this case only the number of robots) and how many experiments need to be conducted for faithfully prediction of the performance in a specific scenario needs currently to be decided on a case-by-case basis.

Also, the proposed modeling building blocks are only approximations of the individual robot behavior (see also Section 4.2.1). Thus "optimal" policies such as the time-varying marker policy (Figure 5.13) might be artifacts of the model. Indeed, an optimal policy for the stationary marker-based algorithm of Section 5.2 could only be found using the collaboration model with deterministic delay (Section 4.2.3) although the system might be equally well represented with the collaboration model based on a probabilistic delay (Section 4.2.3), a model of the inspection delay which would be more in line with those proposed in Section 5.1.3 for the blade inspection duration.

5.5.3   Reactive Algorithms for Robotic Boundary Coverage

Self-organization by positive feedback between the robots showed to significantly improve a non-collaborative, reactive approach for completely inspecting the boundaries of all objects in an environment. This improvement could be shown analytically, numerically, and by real robot experiments. The system is extremely robust despite the limited capabilities of sensors and actuators, and all of the experiments (100 per team size) eventually led to complete coverage. For quantifying this performance in a productive environment, the proposed microscopic models allow for predicting this performance and its distribution.

5.5.4   Multi-Level Modeling

The choice of the abstraction level is a trade-off between the level of detail that needs to be modeled and the experimental time, which one is ready to invest. In practice, initial realistic simulations yield the necessary insight for designing models on higher abstraction levels. These models allow for exploring key parameters relatively easily and yield insight into how to improve the control strategy at the realistic level (see also Figure 4.1, which illustrates this process).

The prediction of macroscopic difference equations and exact microscopic simulation is not always consistent. This is due to the fact that a difference equation integrates, and potentially amplifies, even the smallest fluctuations (resulting for instance from low encountering probabilities), whereas in microscopic simulation events with low probabilities might never actually occur, and therefore do not serve as initial stimulus for other, potentially more likely events. For this reason, results obtained by the macroscopic model should be treated with care, if the average number of agents in a certain state is low. This problem is also illustrated and extensively discussed in (Martinoli et al. 2004).

Emphasis of this chapter is on showing that different modeling abstractions can lead to consistent predictions. However, one abstraction level could be used for studying behavior which is difficult to capture at another level. In particular, the microscopic models allow for exploring aspects of communication/coupling and heterogenities between robots considerably fast, which are potentially not analytically tractable in a straightforward way on the one hand, and unfeasible to implement on the available real robotic platform on the other hand.

5.5.5 Model Complexity

Determining complexity of the model or its state granularity that is necessary for capturing a system's dynamic is a classical problem of System Identification (Johansson 1993). Correll & Martinoli (2006a) show that the dynamics of the inspection can be accurately captured by a model that requires one state less, namely the state `Circle partly inspected element`. This state, however, has been necessary for extending the model to the collaboration case presented in this chapter in order to take into account collaboration between robots that concurrently inspect a virgin blade.

As a rule of thumb, the minimum model complexity is reached, if the model is able to *qualitatively* capture the dynamics of the system. If parameter tuning for *quantitatively* matching an observation leads to counter-intuitive parameter sets in the sense of the proposed parametric calibration method, it is likely that the model misses to capture important properties of the real system, e.g. non-uniform spatial distributions.

5.6 Conclusion

Reactive approaches for coverage have showed to provide competitive performances for distributed coverage, using extremely simple robots. This approach reaches its limitations when the environment is large but the number of robots is small, and if localization is required in order to reconstruct robot trajectories after coverage has been completed, e.g. in an inspection scenario as opposed to a mowing or painting task. Also, reactive coordination coverage is highly redundant, an effect that is not always desirable but can be beneficial, e.g. in an inspection task where the inspection sensor is unreliable.

Using probabilistic population dynamics models is a viable approach for assessing the performance of a reactive, multi-robot system. In particular when the number of robots is large, the system reliably tracks the predicted average system behavior. Probabilistic modeling is not limited to fully reactive systems, i.e. Markovian systems, but can also be applied to semi-Markovian systems with memory as has been illustrated using the marker-based coordination policy. Finally, probabilistic macroscopic models can be used to *optimize* individual parameters of a individual robot's controller and the sequence of the behaviors it executes using dynamic optimization techniques.

Chapter Summary

- A fully reactive policy can lead to robust coverage, which is probabilistically complete and requires only minimialist sensors and actuators.

- Probabilistic, population-based modeling accurately captures the average performance of such a system.

- System identification can be used to achieve quantitatively and qualitatively correct predictions of the system dynamics.

- Optimal control allows for optimizing control parameters of the *individual* robotic platform.

CHAPTER 6

Deliberative Algorithms for Distributed Boundary Coverage

This chapter considers deliberative algorithms on individual and collective robot level. Algorithms are *deliberative* as a robot actively reasons about its actions based on its sensory information and available knowledge about the environment.

The distributed boundary coverage problem is addressed by a suite of algorithms that incrementally raise the assumptions on the robotic platform while increasing the expected performance. In all algorithms in this chapter, boundary coverage is treated as a graph coverage problem (Section 6.1) and algorithms can thus be well applied to distributed coverage problems in general.

First, an algorithm is introduced in which every robot computes individual paths that will lead to complete coverage of the environment in an on-line fashion (Section 6.2). The size and topology of the environment is unknown, and the robots do not collaborate explicitly and their performances solely benefit from parallel task execution. This algorithm is extended by sharing information about task progress among the robots (Section 6.3). This requires communication and the capability to uniquely identify partitions of the environment that a robot has covered. Finally, this chapter introduces a near-optimal partitioning algorithm for the coverage of environments that are known beforehand (Section 6.4). Robots partition the environment using a fully distributed market-based algorithm. In order to be robust against sensor and actuator noise, robots continuously re-plan their trajectories upon reception of new information using the same market-based algorithm that was used for initial partitioning.

Although all algorithms are to a large part deterministic, the fact that they deal with real miniature robotic systems still requires thinking in a probabilistic way. Sensor and actuator noise might lead to the failure of potentially complete algorithms, and thus lead to probabilistic completeness. Similar to the reactive algorithms from Chapter 5, probabilistic microscopic, and where applicable macroscopic, models are used for capturing the actual performance of the deliberative algorithms considered in this chapter. It is then shown experimentally and using probabilistic models that the suite of algorithms presented in this chapter will gracefully decay to the performance of a reactive algorithm.

6.1   Preliminaries

This section describes the environmental models used throughout this chapter, cost functions and team objectives, as well as the reactive robot's behavior, which is used to execute deliberative, high-level control. The individual robot's behavior is also the source of the non-deterministic behavior of the overall system. Model parameters used throughout this Chapter are summarized in Table 6.1.

6.1.1   Environment Model

The cellular decomposition of the environment is described as a undirected graph $G = (\mathcal{V}, \mathcal{E})$ with vertices $\mathcal{V}$ and edges $\mathcal{E}$ (Correll & Martinoli 2007b). Edges represent navigable routes between vertices and can be traversed by a robot in either direction. An individual vertex is denoted by $v \subseteq \mathcal{V}$, and $\mathcal{E}^v = \{e_1^v, \ldots, e_{n_v}^v\} \subseteq \mathcal{E}$ the set of n_v edges incident to v. Obstacles and borders are also represented in $\mathcal{V}$ with a vertex and are identified as non-navigable upon exploring the edge leading to it. The *neighborhood* $\Omega(v)$ of v is given by those vertices that are connected by an edge to v. When on v, a robot can determine all edges incident in v. A sample environment with an arbitrary cellular decomposition is depicted in Figure 6.1.

For each time $t > 0$, a robot i's state is described by $\mathcal{V}_t^i \times \mathcal{E}_t^i \subset \mathcal{V} \times \mathcal{E}$. Thus one robot's knowledge about its environment at a certain point in time is the subgraph $G_t^i = \mathcal{V}_t^i \times \mathcal{E}_t^i$ of G. After reaching and covering vertex v, assuming the average time to do so is τ^v, the robot's state is updated with

$$
\begin{aligned}
\mathcal{V}_{t+\tau^v}^i &= \mathcal{V}_t^i \cup \{v\} \\
\mathcal{E}_{t+\tau^v}^i &= \mathcal{E}_t^i \cup \mathcal{E}^v.
\end{aligned}
\qquad (6.1)
$$

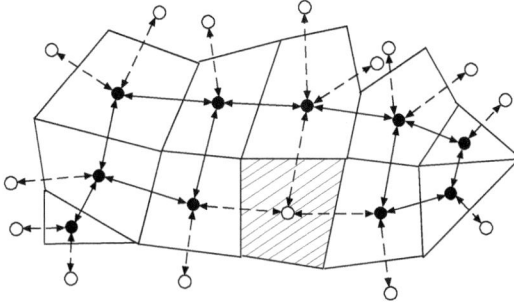

Figure 6.1: An arbitrary environment with a cellular decomposition $\mathcal{V} \times \mathcal{E}$ consisting of a set of vertices $\mathcal{V}$ and a set of edges $\mathcal{E}$. Dashed edges can only be partially navigated and dashed cells are obstacles.

A *partition* of the environment for robot i is given by

$$\mathcal{A}^i \in \mathcal{P}(\mathcal{V}) \tag{6.2}$$

where $\mathcal{P}(\mathcal{V})$ is the power set of $\mathcal{V}$, i.e. the set of all sub-sets of $\mathcal{V}$. A *complete partitioning* $\mathcal{A} \subset \mathcal{P}(\mathcal{V})$ of the environment is given by

$$\mathcal{A} = \{\mathcal{A}^1, \ldots, \mathcal{A}^{N_0}\} = \{\mathcal{A} \mid \cup_{i \in \{1 \ldots N_0\}} \mathcal{A}^i = \mathcal{V}\} \tag{6.3}$$

with N_0 the number of robots. Finally, an *optimal partitioning* $\mathcal{A}^* \subset \mathcal{P}(\mathcal{V})$ of the environment is given by

$$\mathcal{A}^* = \{\mathcal{A} \mid \cup_{i \in \{1 \ldots N_0\}} \mathcal{A}^i = \mathcal{V} \wedge \cap_{i \in \{1 \ldots N_0\}} \mathcal{A}^i = \emptyset\} \tag{6.4}$$

When working without explicit collaboration, it is possible that the environment is effectively covered *before* the individual robots complete, which is given by

$$\bigcup_{i \in \{1 \ldots N_0\}} \mathcal{V}^i_{t_{complete}} = V. \tag{6.5}$$

Throughout this chapter, the definition of $t_{complete}$ from (6.5) is used when referring to *time to completion*.

6.1.2 Cost Functions and Team Objective

The cost that is proportional to the time a robot needs to move from its current position at vertex v to vertex x is defined by

$$C(v, x) \tag{6.6}$$

The cost of the shortest path that connects all vertices of an allocation $\mathcal{A}_i = \{a_1, \ldots, a_n\}$ is denoted by

$$SP(\mathcal{A}_i) = \min_{\pi(\mathcal{A}_i)} \sum_{j=1}^{n-1} C(a_i, a_{i+1}) \qquad (6.7)$$

where $\pi(\mathcal{A}_i)$ is a permutation of $\mathcal{A}_i$. In words, the optimization problem finds the ordering of visits that lead to the shortest overall path. Calculating $SP(\mathcal{A}_i)$ such that every vertex in $\mathcal{A}_i$ is visited only once is an instance of the Traveling-Salesman-Problem[1] (TSP).

The team objective can then be formulated as

$$\min_{\mathcal{A}} \max_i SP(A_i) \qquad (6.8)$$

or in words, finding the partitioning $\mathcal{A}$ that minimizes the longest path a robot has to travel. This team objective is also referred to as MINMAX (Lagoudakis et al. 2005) and is $\mathcal{NP}$-hard (Andersson & Sandholm 2000). Thus the challenge is to find a suitable trade-off between solution quality (efficiency), computation and communication cost (tractability), and the capabilities of the individual robotic platform (feasibility).

Although an optimal partitioning $\mathcal{A}$ is not redundant, the resulting optimal coverage path might well be as robots are free to choose the path between two vertices they want to cover. If no redundancy is required, this would need to be encoded in C, but might be unfeasible in some cases (e.g., in scenarios containing a vertex that has only one edge leading to it).

6.1.3   Reactive Robot Behavior

Exploiting the regularity of the environment for navigation by counting every traversed blade, the *Alice* can construct a graph with the blades as vertices, and possible routes between a blade and its 4-neighborhood as edges (Figure 6.3, *right*, for an example graph). Edge traversal is achieved by a combination of dead reckoning and navigation along way points on a blade's boundary, (Figure 6.3, *left*). At the same time, collisions are avoided reactively.

The algorithms considered in this chapter require the following low-level behaviors: obstacle avoidance, wall following, assessing an objects type (blade, arena boundary, or another robot), determining the blade's type (rotor or stator) at the drop-off location or reading a blade's identification number, navigating to one of two distinct way-points on a blade's boundary, traversing 8 possible edges (4 for rotor and 4 for stator blades), and finally backing up non-navigable edges (i.e. those ending in a wall). The deliberative

---

[1]The exact definition of the TSP requires the traveler to return to the origin

algorithms then sequentially activate the appropriate behavior for moving within the environment. The flow-chart of the robot's controllers is summarized in Figure 6.2. All behavioral parameters are hand-tuned.

Obstacle Avoidance: Obstacles are avoided reactively by calculating the speeds of the left and the right wheel as a weighted sum of the distance measurements (Braitenberg 1986). Additionally, if an obstacle cannot be avoided within a certain time, the robot drives backward.

Wall following: Blades are always circumnavigated in clockwise direction in order to distinguish between blade types, and to be able to use round and sharp tip as reference points for navigation. Wall-following is implemented by a sliding-mode controller which follows the blade's contour using a PD-controller, and performs sharp turns at the tips. Failure of this behavior, i.e. loosing the blade, can be detected if the PD-controller's output is saturated for a certain time.

Assessing an object's type: For distinguishing between the arena border and a blade *without the camera*, the robot exploits the fact that black surfaces hardly reflect infrared light, which lets the robot appear to be farer away from the obstacle then it is. If the obstacle cannot be reached within one second, it is classified as wall. Algorithms using the camera distinguish between walls and blades based on a blade's identification number(walls do not provide an id).

Determining the blade's type: For determining the type of a blade (rotor or stator) *without the camera*, the curvature of the blade between its round and its sharp tip is measured. This is necessary as the required behaviors for navigating to a neighboring blade are a function of its shape. A distinction between the two types can be achieved by counting the number of increments of the wheels's stepper motors. The round and the sharp tip can be distinguished by the amount of sharp turns necessary for surrounding them. In order to reach a certain level of confidence, a robot might need to circumnavigate a blade multiple times. For instance, for determining the blade's type, the difference of "votes" for either type needs to be equal to two, whereby a vote is based on a certain threshold. Parameters determining the termination criteria (achieving a certain sequence of sensor readings) for the behavioral algorithms have been determined experimentally, and aim at a trade-off between accuracy and time needed. Using the camera, the blade type is stored in a look-up table together with its identification number.

Way-point navigation: In order to reach blades in its 4-neighborhood, the robot relies on two way-points that are detectable by the robot's on-board sensors (Figure 6.3). The robot follows the blade contour until it either passed the round tip (way-point ❶) or the sharp tip (way-point ❷). It distinguishes between round and sharp tip by the number of sharp turns needed by the sliding-mode controller for wall following. Unlike the behavior in Chapter 5 which let robots translate along a blade from sharp tip to round end in open-loop control, in this Chapter robots actively try to estimate their location on the blade by processing sensory information.

Edge Traversal: In order to translate to another blade, the robot first turns on the spot in the right direction, and then executes a forward motion until an obstacle is reached. Hereby, obstacles encountered until a certain time-out are disregarded, allowing the robot to move close to the blade's contour, which is necessary to reach some configurations. Parameters such as the turn angle, the wheel speeds for the left and right wheels, and the time-out are hard-coded in the extension module and send to the Alice. Whether traversal failed can only be determined by the *Assessing an object's type* behavior and only if the topology of the environment is already known.

Backup: The robot moves backwards until it encounters an obstacle (open-loop control).

Navigation error

The reactive behaviors described above potentially fail due to sensor and actuator noise (in particular wheel-slip on the Alice platform). While some closed-loop behaviors are sufficiently stable (e.g., wall-following), open-loop behaviors such as navigating from blade to blade might fail completely. Such a situation can be detected on the deliberative level: for instance when the robot encounters a situation where its sensor readings do not match what it expects, e.g. when it encounters a wall where there should be a blade. Other reasons for failing are expiration of time-outs when trying to attaching to a blade or determining an objects type. When no global reference is available the deliberative controller is "lost" at this point and needs to start over. On the other hand, if a global reference is available, such as global localization, here given by color codes on the blades, the deliberative algorithm can use this information to recover.

Depending on the available hardware, the following additional behaviors upon a detected failure are implemented. If no localization is available, a robot starts over at its current location after performing a short random walk (2s). If localization is available, the robot always uses the current position as basis for path-planning. Thus, if a robot does not fail completely (battery failure, being mechanical stuck) it will eventually arrive at a blade and plan its motion from there.

6.1.4 Microscopic Simulation

The environment and the robotic platform are simulated in the realistic, sensor-based simulator *Webots* (Section 3.3). Webots allows for varying sensor and actuator noise

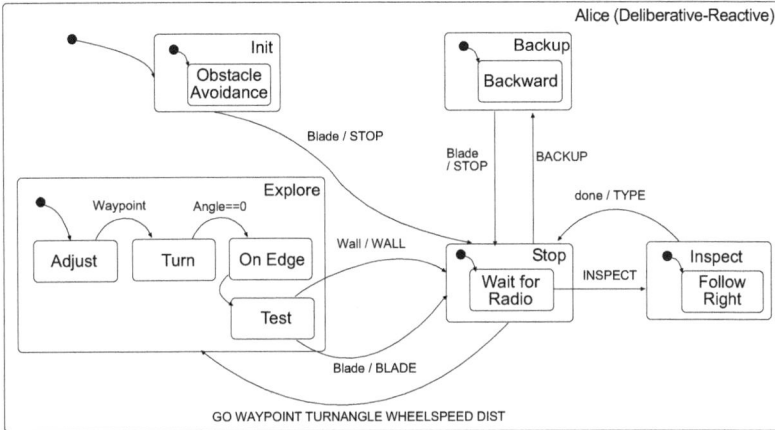

Figure 6.2: Stateflow diagram of the robot controller implementing the behavioral layer of the deliberative-reactive algorithm. Black dots denote initial states. Boxes denote the encapsulation of behaviors.

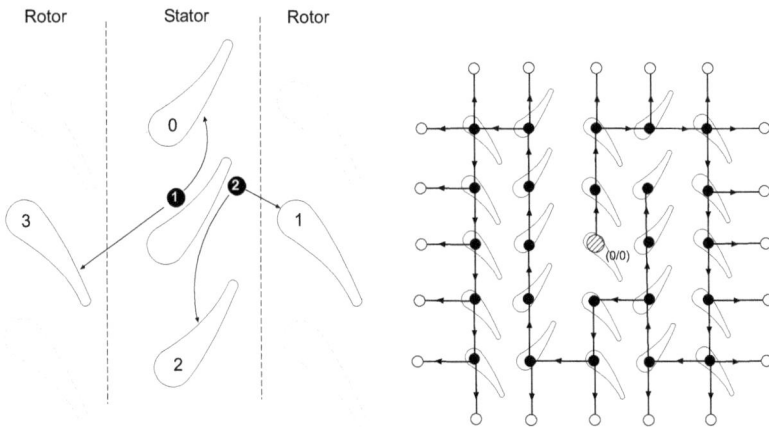

Figure 6.3: *Left:* Way-points on a blade's boundary that can be navigated to using on-board sensors. *Right:* Possible trajectory for a single robot along a spanning-tree in a 5x5 blade environment (bold line). Backtracking paths are not shown.

parameters of the individual platform and for extracting probability distributions for the reliability of an individual robotic platform.

These probability distributions can then be used to simulate the system. Unlike the example given in Section 4.3.1 all possible system states are not explicitly enumerated but the environmental state is given by the graph structure introduced for the particular problem of this Chapter in Section 6.1.1 and the robot state given by the subgraphs defined in (6.1). Possible trajectories through state space are then simulated by updating the robot states with a combination of the microscopic control law or randomly moving it according to the probability distribution describing the robot's reliability.

As also the duration for different behaviors required by one step of the deterministic algorithm follow a distribution (see the Example in Section 4.3.2) and the time required for inspecting one blade is long compared with the time-discretization that would be needed for accurately modeling the rather wide distribution of inspection time, all algorithms are simulated by discrete event system (DES) simulators which allows considerable speed-up when compared with a synchronous simulation[2].

In the DES simulator, unlike in realistic simulation, multiply point-robots can occupy one vertex, without penalty. One simulation step consists in moving to a vertex, determining the vertex to visit next and potentially communicating/bidding for state information. If applicable the direction of motion and duration of a specific state are probabilistic.

6.2   Non-Collaborative Deliberative Coverage without Localization

If no absolute localization mechanism is available, robots can only determine their relative position with respect to their drop-off position by tracking the traversed vertices and create a topological map in an on-line fashion. As the topological maps created by such a method are potentially ambiguous without a global reference frame, robots do not exchange information when no localization is available.

The algorithm approximates the team objective (6.8) by constructing independent partitions for each robot until the termination criterion (6.5) is reached. As robots do not collaborate, individual robots are unaware of task completion.

---

[2]Non-collaborative and collaborative algorithms for unknown environments have been implemented in MATLAB, whereas market-based algorithms have been implemented in Java for performance reasons

Environmental states	Description	Section
$G = (\mathcal{V}, \mathcal{E})$	The environment is modeled as a graph consisting of vertices $\mathcal{V}$ and edges $\mathcal{E}$.	6.1.1
v	An individual vertex $v \subseteq \mathcal{V}$	6.1.1
$\mathcal{E}^v$	The set of edges incident to v	6.1.1
$\Omega(v)$	The set of edges connected to v by not more than one edge	6.1.1
Robot states		
$\mathcal{V}_t^i, \mathcal{E}_t^i$	The set of vertices and edges known to robot i at time t	6.1.1
$\mathcal{A}^i$	Set of vertices (partitioning) assigned to robot i	6.1.1
$\mathcal{A}$	Set of partitionings over all robots	6.1.1
System parameters		
M_0	Number of blades	5.1.2
N_0	Number of robots	5.1.2
τ^v	Time to traverse an edge and cover a vertex	6.1.1
π^e	Probability to successfully traverse an edge	6.2.2
μ	Average number of vertices covered before failure ($\frac{1}{1-\pi^e}$)	6.2.2
π_f	Probability of erroneously localization	6.5.2

Table 6.1: Summary of model parameters used in this chapter with the section in which they have been introduced.

6.2.1 Individual Robot Behavior

The robot explores the environment by moving from vertex to vertex using the reactive behaviors described in Section 6.1.3 using a greedy algorithm that moves towards the closest unexplored edge. Exploiting the information that every vertex but the walls has 4 neighbors and vertices are aligned in a grid in this case study, the robot creates a topological map of the environment with up to 3 unexplored edges per vertex. After all vertices have been visited, the robot starts over. For calculating the cost function C, Dijkstra's algorithm with an uniform edge weight of 1 is used. If there is more than one x which minimizes C, one of them is chosen at random.

If a behavior fails or if there is a mismatch between the coverage map and the actual robot position (e.g., a robot arrives at a wall where should be a blade), the coverage map is erased and the robot starts over from scratch (empty coverage map).

6.2.2 Microscopic Model

A robot picks the closest vertex from the set of discovered, reachable and not visited vertices DRV by solving

$$\min_{x \in DRV_t^i} C(v, x) \tag{6.9}$$

The information obtained while calculating $C(v, x)$ can then be used for determining the best next vertex $v' \in \Omega(v)$ to visit on the path towards x.

The set of discovered, reachable and not visited vertices is defined by the intersection of the set of vertices RV that are reachable from the current robot's location with the set of discovered, but not visited vertices DV:

$$DRV_t^i = RV_t^i \cap DV_t^i \tag{6.10}$$

Setting $C(v, x) = \infty$ if $x \notin RV_t^i$, (6.10) can be simplified to $DRV_t^i = DV_t^i$. The sets DV and RV can be calculated given the robot state $\mathcal{V}_t^i \times \mathcal{E}_t^i$ as follows. The discovered but not yet visited vertices DV, are the vertices where one and only one vertex of a known edge is a visited vertex, i.e.

$$DV_t^i = \{v \in \mathcal{V} \mid (v^*, v) \in \mathcal{E}_t^i \wedge v^* \in V_t^i \wedge v \notin V_t^i\} \tag{6.11}$$

where (v^*, v) is the edge connecting v^* and v. The reachable nodes with respect to vertex v are given by the following recurrence relation:

$$(RV_t^i)_j = \{v^* \in \mathcal{V} \mid (v, v^*) \in \mathcal{E}_t^i \wedge v \in (RV_t^i)_{j-1}\} \tag{6.12}$$

with $(RV_t^i)_1 = \{v\}$. In words, $(RV_t^i)_j$ contains all vertices that are reachable from v by at most j edges.

Modeling Navigation Error

For accounting for potential robot failure, a *probability* is associated with every edge traversal. With the probability π^e to successfully traverse an edge e, the probability to reach and cover vertex x can be calculated by the following recurrence equation

$$\pi_{i,x}(t + \tau^x) = \pi_{i,x}(t)\pi^e \tag{6.13}$$

where e is given by (v, x) with x being the next vertex according to (6.9).

If π^e is constant for all edges, and Markovian, that is only dependent on the robot's current state, the probability to fail after traversing n edges (i.e. it failed on the nth edge) follows a geometric distribution

$$P_{geo}(n) = (1 - \pi^e)(\pi^e)^{n-1} \tag{6.14}$$

and the average number of vertices before failure μ calculates[3] to $\mu \approx \frac{1}{1-\pi^e}$.

For microscopic simulation, navigation error occurs with probability $1 - \pi^e$ when moving from vertex to vertex. Then a robot is placed at a random location and its coverage map is erased.

6.2.3   Macroscopic Model

An exact macroscopic model would require master equations for every possible partition-ing of the environment (see Section 4.3.1), which becomes untractable already for small environments. A possible approximation assumes that every distribution of robots in the environment is equally likely and independent of coverage progress. With the same probability to fail for each robot, the following recurrence equation for the average num-ber of uncovered vertices after κ trials can be written (Correll & Martinoli 2007b). One trial is considered to be the coverage of μ distinct vertices that are part of a spanning tree constructed by a robot using (6.9) before it does a mistake.

$$M_v(\kappa + 1) = M_v(\kappa) \left(1 - \frac{\mu}{\|\mathcal{V}\|}\right)^{N_0} \tag{6.15}$$

The duration of one trial is given by $\mu\tau^v$, the average time needed for covering μ vertices. Equation (6.15) has a similar form as the model in Chapter 5, equation (5.7), where a probabilistic model for random coverage of the environment is proposed: the likelihood of covering a virgin vertex decreases exponentially with the number of already covered vertices.

---

[3]This is an approximation as the graphs usually have finite size.

When $\mu = \|\mathcal{V}\|$, that is the whole graph is covered in one trial, (6.15) predicts exactly the same time to completion as the model for ideal robots ($\|\mathcal{V}\|\tau^e$). On the other hand, for $\mu = 1$, that is robots are unable to enforce the deliberative control policy, a robot is moving essentially randomly.

6.2.4 Completeness

The policy from (6.9) will lead to complete coverage of the environment as the robot will always move towards remaining unvisited vertices. Sensor and actuator noise, however, leads to asymptotic coverage and convergence towards complete coverage follows from (6.15) as $M_v(k)$ is monotonically decreasing for $\frac{\mu}{\|\mathcal{V}\|} \leq 1$.

6.3 Collaborative Deliberative Coverage with Localization

If localization and communication are available, robots can share information about coverage progress using a wireless link. Using these additional assumptions, the algorithm presented in this section leads to complete coverage of the environment with improved time to completion when compared with the non-collaborative algorithm from Section 6.2. As before, errors that are typically encountered in real miniature robotic systems are explicitly taken into account. In addition to navigation error introduced by wheel-slip (Section 6.2.2), the effect of erroneous localization information on the coverage performance is studied experimentally and analytically. The presented distributed coverage algorithm is robust towards localization errors and to unreliable communication. Also, it allows for calculating lower bounds for the probability to complete.

The algorithm approximates the team objective (6.8) by constructing partitions for each robot until the termination criterion (6.5) is reached. Due to communication among the robots, the resulting partitioning is optimal in the sense of (6.4), but does not necessarily minimize (6.8) as the resulting trajectories are potentially sub-optimal as robots race for uncovered vertices instead of arbitrating them among each other.

6.3.1 Individual Robot Behavior

A robot constructs a topological map of the environment as in Section 6.2.1. Every time a robot extends its map, the map is broadcasted via the radio. As new information due to topological maps received from team-mates potentially render the current goal of a robot obsolete, the goal is re-computed every time an edge is traversed. This policy also leads to robustness against navigation errors as the next direction to move is always calculated with respect to the current position. A robot shares his complete state, which minimizes the potential loss of information due to communication faults.

As the localization mechanism might lead to an erroneous belief about the robot's position, robots perform multiple tours in order to increase the likelihood for complete coverage. The number of times a vertex has been visited is stored together with the topological map and is also broadcasted via the radio. Instead of moving to the closest unvisited vertex as in Section 6.2.1, the robot moves now towards the closest vertex with the lowest tour index. The algorithm terminates after all vertices have been visited for a certain number of times.

6.3.2 Microscopic Model

After robot j visits a vertex, it will update its state G_t^j according to (6.1) and then broadcast it. When robot i receives the coverage map G_t^j from robot j, it will update its own state by merging the two maps, that is

$$G_{t+\epsilon}^i = G_t^i \cup G_t^j \tag{6.16}$$

where ϵ is the necessary time for communication and information processing. A robot needs to evaluate (6.9) each time new information is available as the current goal vertex might already have been visited by another robot.

A tour index $\mathcal{T}^i(v)$ is associated with each vertex $v \in \mathcal{V}$. Upon visit of a vertex v, $\mathcal{T}^i(v)$ is increased by 1. Initially, $\forall v \in \mathcal{V}, \mathcal{T}^i(v) = 0$.

The set of vertices a robot should visit next is then given by

$$\begin{aligned} DV_t^i &= \{v \in \mathcal{V} \mid (v^*, v) \in \mathcal{E}_t^i \wedge \\ &\quad v^* \in V_t^i \wedge (v \notin V_t^i \vee \mathcal{T}^i(v) = \mathcal{T}_{min})\} \end{aligned} \tag{6.17}$$

which will direct a robot to visit a vertex with the lowest tour index, given by $\mathcal{T}_{min}^i = \min_{v \in \mathcal{V}_t^i} \mathcal{T}^i(v)$. At the same time (6.17) ensures that robots will complete one tour before beginning the next one. For all $v \in \mathcal{V}_t^j$, $\mathcal{T}^j(v)$ is broadcasted along the information mentioned in Section 6.3.1. Upon receipt, robot i will additionally update its state as follows.

$$\mathcal{T}^i(v) = \max(\mathcal{T}^j(v), \mathcal{T}^i(v)) \tag{6.18}$$

With probability $1 - \pi^e$ a navigation error occurs, and the robot is positioned on $\Omega(v')$, i.e. in the neighborhood of the vertex it wants to reach. With probability π_f the belief on the robot's position is chosen randomly from $\mathcal{V}$.

6.3.3   Macroscopic Model

An exact macroscopic model would require master equations for every possible partition-ing of the environment (see Section 4.3.1), which becomes untractable already for small environments. Available localization which provides a global reference frame enables the robots to recover from navigation errors. Thus, unlike in Section 6.2.3, the likelihood for a certain distribution of the robots in the environment is a function of coverage progress (robots tend to move towards unexplored regions), and (6.15) becomes a lower bound for performance.

The number of tours required for achieving a certain confidence in coverage can be calculated as follows. Given a probability π_f for incorrect localization,

$$m \geq \lceil \frac{\ln \alpha}{\ln \pi_f} \rceil \qquad\qquad (6.19)$$

tours are required to achieve an average coverage level of $1 - \alpha$. The probability of erroneously localizing itself in m independent trials is given by $\pi_f{}^m$, which leads to the inequality $\pi_f{}^m \geq \alpha$ for the probability α to fail on an individual cell.

Assuming π_f to be constant over the whole environment, α can also be understood as the fraction of cells where localization failed. For example, in order to achieve coverage of 95% given a sensor that provides correct localization with a probability of $\pi_f = 70\%$ requires $m = 3$ tours. m is an upper bound because (1) our analysis does not take into account the cells which are visited at the place of the cell the robot believes it visits, and (2) we do not take into account that the robot will redundantly visit cells while on the way to an unvisited cell.

6.3.4   Completeness

Upon availability of global localization, navigation error as described in Section 6.2.2 only lead to a delay in algorithmic execution but not to complete failure. In this case, the algorithm described in Section 6.2.2 leads to complete coverage if at least one robot does not fail.

Theorem 6.3.1. Complete Coverage for a single robot: *Coverage is completed, when* $\min_{x \in DRV_t^i} \mathbf{C}(v, x) = \emptyset$ *(6.9), that is there are no discovered, unvisited vertices ($DRV_t^i = \emptyset$) and all vertices are reachable ($RV_t^i = V$).*

Proof. Using (6.10), DRV_t^i is an empty set, when all discovered vertices have been visited ($DV_t^i = \emptyset$). Using (6.11) this is the case if there exist no edge that connects an unvisited to a visited vertex. When there are no edges with unvisited vertices, $RV_t^i = V$.   □

Completeness for a multi-robot team follows directly from the proof for a single robot. Completeness for noise in localization is asymptotic. As increasing the number of tours m in (6.19), α steadily decreases, and thus the algorithm converges to complete coverage.

6.4 Market-Based Distributed Coverage

This section considers distributed coordination policy for covering an environment that is known in advance. In addition, every robot needs an unique identification number. Knowing the environment before-hand has the advantage that potentially optimal trajectories for every robot can be calculated off-line.

A promising approach for finding a near-optimal partitioning are so called *market-based algorithms* (Dias, Zlot, Kalra & Stentz 2006). Market-based algorithms have been specifically used for multi-robot coverage by (Rekleitis et al. 2005) and exploration (Zlot & Stentz 2006). Formulated as distributed coverage problem, trading goods correspond to cells of a cellular decomposition of the environment, and prices correspond to the cost (energy consumption, e.g.) for an individual robot or the benefit to a particular team objective. The algorithm presented in this section bases on a market-based solution approach known as *Free Market* introduced by Dias & Stentz (2000). *Free Market* represents a market in which trade is determined by unregulated interchange, i.e. not under the control of a centralized auctioning entity. This market-based approach has been chosen for two reasons. First, as no centralized control over auction is needed, it allows for a decentralized implementation on real robots. Second, Lagoudakis et al. (2005) provides upper and lower bounds on the performance of this approach.

The algorithm approximates the team objective (6.8) by constructing an optimal partitioning. Unlike the algorithms from Sections 6.2 and 6.3, the bidding process aims explicitly at minimizing (6.8) by solving the TSP for every partitioning under auction. Even for an optimal solution to the TSP, this approach cannot guarantee an optimal solution, as the partitioning is constructed iteratively, i.e. one vertex after the other is assigned to the robot with the smallest cost (as opposed to combinatorial auctions, see Berhault et al. 2003).

6.4.1 Individual Robot Behavior

Initially, all tasks are un-allocated and robots are randomly walking in the environment. As soon as a robot encounters a blade it reads its position and waits until every other robot has reached a blade, which is considered its initial position, or a time-out occurs (in case some robots fail in this initial phase). Alternatively, robots can be positioned at a central location. Bidding is then conducted round-wise. Each robot calculates bids on all unallocated vertices and send its "best" bid, i.e. the bid which is most likely to succeed, to the other robots.

Bids are based on locally available information, i.e. a robot's partial allocated set of vertices $\mathcal{A}_i$ and its position. The robot with the winning bid for a vertex gets this vertex assigned. As there is no central auctioneer, bid selection is performed locally on each robot and each robot i updates $\mathcal{A}_i$ and the set of un-allocated vertices. Thus, every robot itself plays the role of the auctioneer and associates vertices with robots and only the computation of bids is distributed among the robots. This approach has the advantages that there is no central auctioneer that could potentially fail but also that every robot maintains its own view on the availability of other robots and assigns their tasks to itself in worst case (e.g. due to communication failure).

Bids are calculated as follows. With $\mathcal{A}_i$ the current partition assigned to robot i, x_i its current position, and v an unallocated vertex, robot i bids the total cost for its current allocation including the unallocated vertex $SP(\mathcal{A}_i \cup x_i \cup v)$.

As all bids are known to all robots, each robot can locally determine the winning bid. In order to yield the same outcome for each robot, determination of the winning bid needs to be fully deterministic. First the vertex which can be associated with the lowest cost is determined. This vertex is allocated to the robot providing the lowest total cost for this vertex. If two or more robots offer the same lowest total cost, the vertex is allocated to the robot providing the lowest marginal cost, i.e. $SP(\mathcal{A}_i \cup v) - SP(\mathcal{A}_i)$. If two or more robots offer the same lowest total cost and lowest marginal cost, the vertex is allocated to the robot with the lowest identification number. Thus, at each round of the auction one vertex will be assigned to a particular robot.

Bidding is terminated after all vertices have been allocated, i.e. after M_0 rounds. An example (optimal) partitioning for the environment of the turbine inspection case study is shown in Figure 6.4, *left*. Coverage is terminated after every robot has completed all tasks it was assigned to.

$SP(\mathcal{A}_i)$ is calculated using a near-optimal constructive heuristic for the TSP (see (Amstutz 2007) for details on the implementation).

Robustness towards sensor and actuator noise

Due to sensor and actuator noise it is likely that a navigation error occurs. Also, the non-deterministic time for navigating from vertex to vertex (see Figure 6.6) might lead to sub-optimal execution times, e.g., when finished robots could take over vertices from a robot that has been slowed down. A scenario showing recovery from a navigation error is depicted in Figure 6.4, *right*. The algorithm described in Section 6.4 is thus extended by the following additional rules (Amstutz 2007):

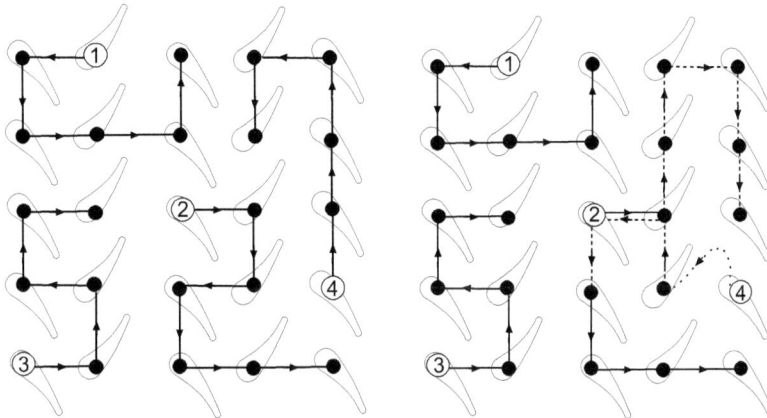

Figure 6.4: *Left:* An optimal partitioning for four robots at random drop-off locations. *Right:* Robot four failed traversing the first edge and trajectories are re-auctioned, leading to a (less optimal) task allocation for robot 2 and 4 (dashed lines).

- After every successful coverage, the robot broadcasts this information to the team and calls an auction using the same scheme as for the initial partitioning. This time, however, the auction is limited to the current set of uncovered vertices.

- Whenever a robot crossed an edge on the shortest path spanning $\mathcal{A}_i$, it verifies whether navigation was successful. If this is not the case, it calls an auction as above.

If a robot does not take part in an auction, e.g., due to packet loss or robot failure, it is not considered in the partitioning, which potentially leads to redundant coverage but retains the completeness properties of the algorithm. Also, if an auction ends but a robot did not receive any bids for some vertices, the robot adds those vertices to its partition.

6.4.2 Microscopic Model

With $\mathcal{A}_t^i = \{a_1^i, \ldots, a_m^i\}$ the set of remaining vertices of robot i at time t and a_0^i its current position, a robot calculates the shortest path that connects all remaining vertices by

$$\pi(a_1^{i*}, \ldots, a_m^{i*}) = \arg \min_{\pi(\mathcal{A}_t^i \cup a_0^i)} \sum_{j=0}^{n-1} C(a_i, a_{i+1}) \qquad (6.20)$$

Thus $\pi(a_1^{i*}, \ldots, a_m^{i*})$ corresponds to the sequence of vertices that are connected by the shortest possible path with respect to a robots current location a_0^i.

A robot moves along $\pi(a_1^{i*}, \ldots, a_m^{i*})$ where the next vertex v' is calculated by solving $\mathcal{C}(a_0^{i*}, a_1^{i*})$. With probability $1 - \pi^e$ a navigation error occurs, and the robot is positioned on $\Omega(v')$, i.e. in the neighborhood of the vertex it wants to reach. With probability π_f the belief on the robot's position is chosen randomly from $\mathcal{V}$.

$\mathcal{A}_t^i \subseteq \mathcal{A}_t$ corresponds to the solution of (6.8) using the market-based algorithm described in Section 6.4.1. $\mathcal{A}_t$ is recalculated every time a robot reaches a vertex. By this, potential robot failures such as navigation error, communication loss and complete failure are accommodated.

6.4.3   Completeness

The algorithm is complete as long as at least one robot stays alive. As every auction consists of M_0 rounds and in each round exactly one robot is assigned a vertex, the partitioning of the environment is complete. This holds also for auctions hold for partly covered environments and with incomplete teams. Upon communication failure, the environment will be covered redundantly.

6.5   Results

For experiments requiring communication, a Telos (Polastre et al. 2005) mote serves as repeater and periodically repeats received messages at 10Hz. This allowed for significantly reducing the on-time of the radio-module, which is only turned on when a robot moves to a new vertex or finishes coverage of a vertex. For experiments with market-based algorithms that require extensive computation on the individual node, and as local bid-computation and selection is unfeasible using the limited computational capabilities on the extended Alice platform, bidding and task-allocation was performed off-line on an external computer based on the actual positions of the robots that are communicated by radio. The base-station then performs re-planning upon availability of new information (e.g., robot failure, coverage progress) and periodically broadcasts $\mathcal{A}$ to the robots, which are endowed with a map of the environment.

6.5.1   Non-Collaborative Deliberate Coverage without Localization

We will first validate our assumption that robot failures are Markovian and calibrate model parameters π^e, μ and τ^v. We then use these results to compare the prediction of the deterministic model and the probabilistic model with results from Webots.

Wheel-slip	π^e	μ	$\tau^v[s]$
0%	1	25	12
10%	0.79	4.77	38.5±10.3
50%	0.67	3.03	44±20.5
real	0.64	2.79	52±19.7

Table 6.2: Calibrated model parameters for different amounts of wheel-slip in realistic simulation and real robots.

Calibration of Model Parameters

For validating our assumption that sensor and actuator noise leads to a time-independent (Markovian) probability to violate the completeness properties of a deliberative coverage algorithm when moving from blade to blade when no global localization mechanism is available, we measure the number of blades a single robot can traverse without mismatch between its actual location on the spanning tree and its belief occurring in *Webots* for random drop-off locations and wheel-slip of 10% and 50% (6000 experiments each), as well as for a real robot (50 experiments). From this results (Figure 6.5), we calculate the probability π^e of successfully traversing an edge using (6.14), the average time τ^v an edge traversal takes (including coverage of a vertex), and the average number μ of covered vertices before the completeness properties of the algorithm are violated (Table 6.2). Results for τ^v are shown in Figure 6.6. The distribution his skewed and contains values up to 117s for 10% wheel-slip and 630s for 50% wheel-slip (median 38s and 41s, respectively).

Experimental results

The times it takes to completely circumnavigate every one of 25 blades in the arena of Figure 3.3 for 10 robots and wheel-slip of 10% and 50% (100 experiments per team size and wheel-slip) as well as with a team of 10 robots are shown in Table 6.3 (10 experiments). Table 6.3 also shows the average number of vertices visited by a single robot before complete coverage was achieved.

Model Prediction vs. Realistic Simulation

The average number of covered blades is measured over 100 experiments in *Webots* vs. time. Results for teams of 1 and 10 robots (lower and upper curve, respectively) for wheel-slip of 10% and 50% are shown in Figure 6.7, *left*, and *right*. These results are compared with predictions from the probabilistic model (6.15) for parameters μ from

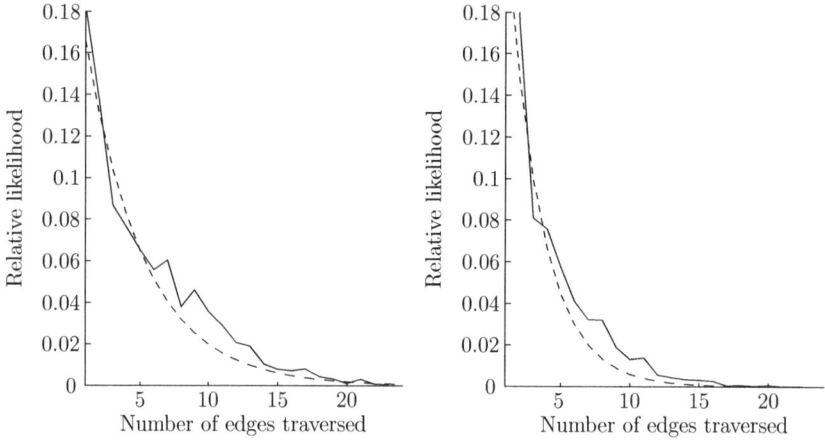

Figure 6.5: The relative likelihood for successfully traversing a certain number of edges for wheel-slip of 10% (*left*) and 50% (*right*) matches a geometric distribution (superimposed).

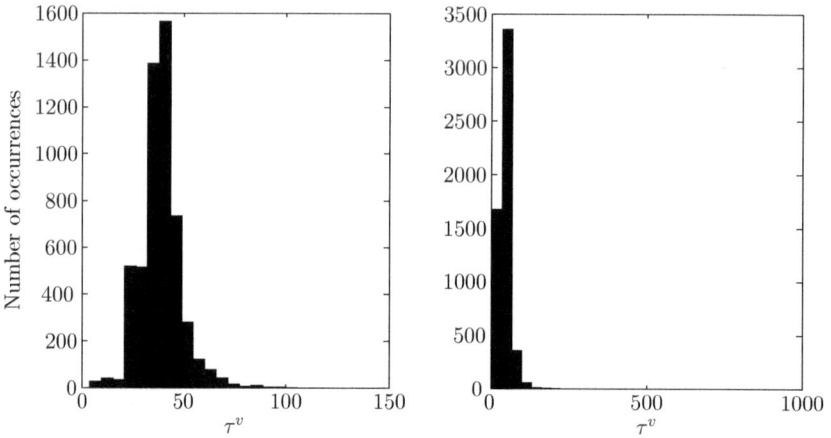

Figure 6.6: Histogram of blade-to-blade navigation and coverage time τ^v for 6000 reactive blade-to-blade transitions with wheel-slip of 10% (*left*) and 50% (*right*).

Wheel-slip	Avg. nb. of vertices visited to completion	Time to completion for 10 robots [s]
0%	25–250	36–300
10%	63.21	541±252
50%	78.58	701±342
real	n.a.	788±375

Table 6.3: Redundancy and time to completion for different amounts of wheel-slip in realistic simulation and real robots.

Table 6.2.

6.5.2 Collaborative Deliberative Coverage with Localization

Localization requires an additional parameter for modeling localization error in the discrete-event simulation that is calibrated based on real-robot experiments. Using this parameter as well as π^e and τ^v, DES and Webots simulations show good agreement among each other and with real robot experiments. Finally the impact of limited range communication and localization error are studied using the probabilistic model. The scaling behavior with respect to the environmental size and the relative improvement by communication vs. non-collaboration are discussed in (Rutishauser et al. 2007).

Calibration of Model Parameters

For testing the visual localization algorithm (Section 3.1.3) the camera module was manually placed in front of a blade and 100 pictures per code (25 codes) were taken. Measurements lead to erroneous localization in $p_f = 5.03\%$ of the cases (see Rutishauser 2007). Although misalignment of the camera with respect to the blade was simulated manually, the measured p_f can be considered as lower bound for the performance in the real setup where changing lighting conditions or interrupt by other robots will lead to even lower detection ratios. The effective p_f is also dependent on the blade geometry as the identification cannot be recorded when the blade is approached from its tip. Simple geometric considerations (Rutishauser 2007) lead to an effective $p_f = 33\%$.

Experimental Results

Realistic simulation from Webots for 1 to 10 robots with and without collaboration, i.e. with and without global communication, are compared to results obtained from

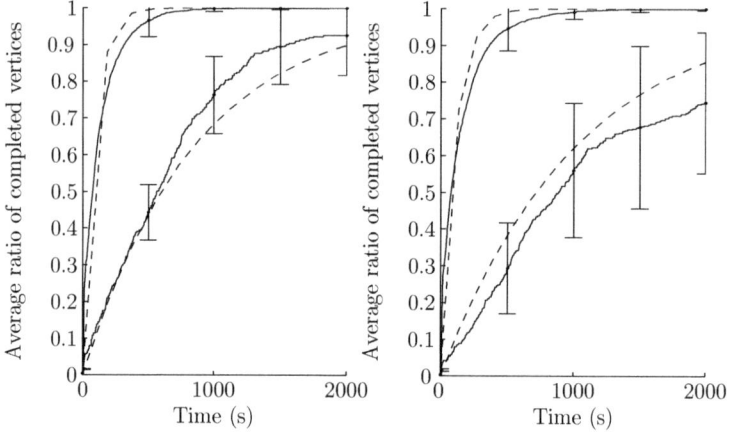

Figure 6.7: Average ratio of covered blades for teams of 1 and 10 robots (coverage progress with 10 robots is faster) and wheel-slip of 10% (*left*) and 50% (*right*). Prediction of the probabilistic model (dashed) is superimposed. Error bars depict standard deviation.

the discrete-event simulation (DES) by the median time to completion (Figure 6.8) and sampling τ_v from Figure 6.6, left. We use the non-parametric Wilcoxon rank-sum test at 95% to determine if the qualitative match in medians between the Webots and Matlab data for different number of robots can be statistically verified. In other words, the probability that data from realistic simulation and discrete-event simulation has a different distribution is less than 5%. Using communication, in nine out of the ten cases we can establish that match. When not using communication, almost half of the cases do match, while still providing close qualitative agreement.

Modeling prediction from Webots and DES are also compared to experiments conducted with 5 and 10 real robots in Figure 6.9 (left bars).

Collaborative vs. Non-Collaborative Coverage

As global communication is a strong assumption that can not always be guaranteed, four representative communication ranges are evaluated using discrete event simulation: no communication (range 0), robots communicate only when on the same vertex v, robots communicate only with robots in their neighborhood $\Gamma(v)$, and global communication, i.e. over $\mathcal{V}$. Time to completion for a 10x10 square lattice are depicted in Figure 6.10. There is already considerable improvement in performance when communication is limited to

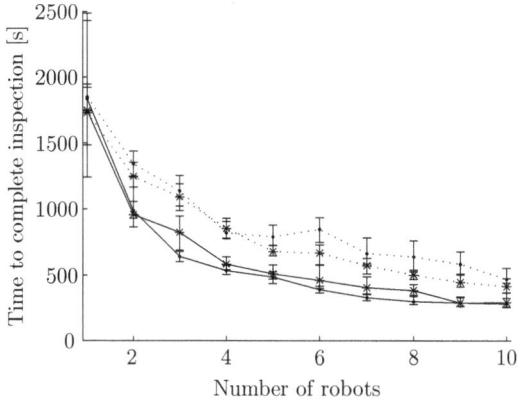

Figure 6.8: Realistic simulation ($*$) and prediction of the discrete-event simulation ($\cdot$) for 1 to 10 robots in the 5x5 environment and 10% wheel-slip for global (——) and no communication ($\cdots$). Both model abstraction levels provide good qualitative and quantitative agreement. 100 replications for both Webots an DES.

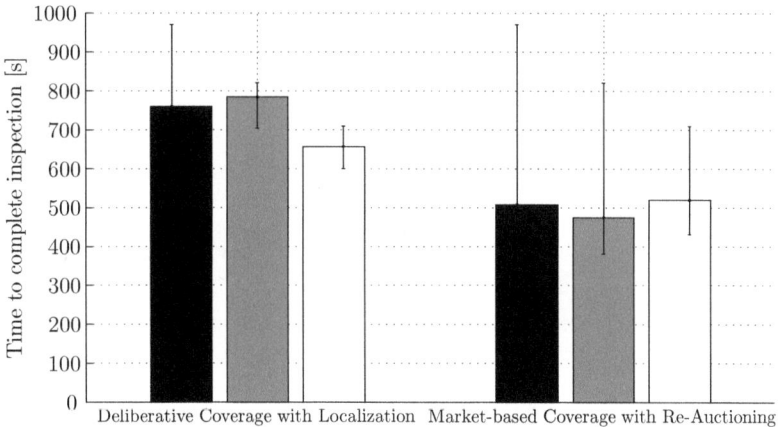

Figure 6.9: Comparison of real robot experiments (■), DES simulation (■), and realistic simulation (□) for 50% wheel-slip and $p_f = 33\%$. 100, 100, and 9 replications for Webots, DES, and real robots respectively.

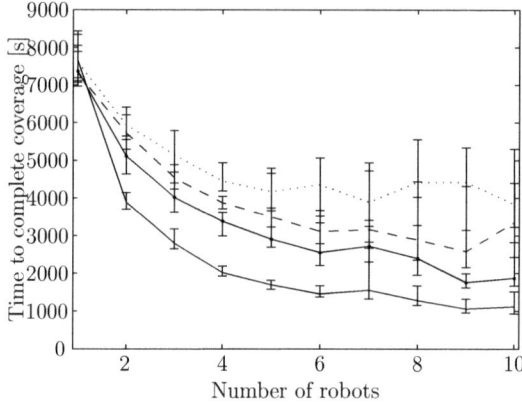

Figure 6.10: Influence of the communication range for 1-10 robots in a 5x5 grid without noise for the collaborative algorithm with localization: no communication ($\cdots$), same vertex ($--$), neighborhood ($-\cdot-$), and global ($—$). Performance is significantly improved as soon as communication is available.

robots being on the same or neighboring vertices.

Imperfect Localization

Imperfect localization was tested on the discrete event simulator for the 5x5 square lattice. On every vertex, a robot was provided with the position of a random vertex with probability π_f. Notice that error in localization not only leads to sub-optimal path planning but also to wrong information shared with other robots. Results comparing no-communication with global communication for different π_f are shown in Fig 6.11. One sees that when π_f increases, the benefits of using communication get smaller. Up to a certain number, which seems to be a function of the environment size, using more robots which communicate obviously improves the results.

In order to validate the upper bound for the number of required tours coverage progress is measured after every robot performed $M = 2$ tours (for $\pi_f = 10\%$ and $\pi_f = 20\%$), to $M = 3$ tours for $\pi_f = 30\%$ and to $M = 4$ tours for $\pi_f = 40\%$ error; the number of tours were calculated using $\alpha = 0.05$. In all simulations, the average coverage is above 99% ($\alpha \leq 0.01$).

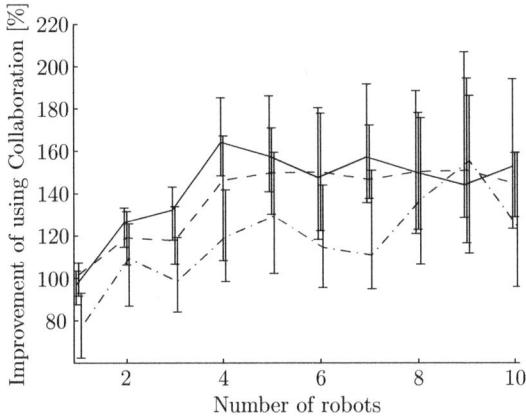

Figure 6.11: Improvement of global communication vs. no communication (100%) for 1 to 10 robots and different level of localization error $p_f = 0\%$ (—), $p_f = 10\%$ (– –), and $p_f = 40\%$ (–·–) in a 5x5 environment. Communication is even beneficial when localization is noisy and becomes increasingly beneficial with larger team sizes.

6.5.3 Market-Based Distributed Coverage

Results for 9 experiments with 5 real Alice robots are shown in Figure 6.9 (right bars). Prediction of the DES simulator and Webots for wheel-slip of 10% and 50% provide close match and performance gracefully decays with increasing amount of slip noise (Figure 6.12), 100 experiments per team-size/noise level in a 5x5 grid.

DES and Webots are also used to study the impact of the communication range. Figure 6.13 shows results for global communication and no communication, i.e. zero range communication. Performance gracefully decays for intermediate levels (Amstutz 2007).

6.6 Discussion

6.6.1 Sensor and Actuator Noise

Varying amounts of wheel-slip in realistic simulation leads to a constant probability for violating the completeness properties of the algorithm without localization. The variance of the performance increases with the wheel-slip (compare Figure 6.7 *left* and *right*). This can be explained by robots behaving less faithfully to their deliberative control scheme with increasing amounts of noise, and thus behaving less predictable. As soon as (perfect) localization is available, robots can recover from navigation error and

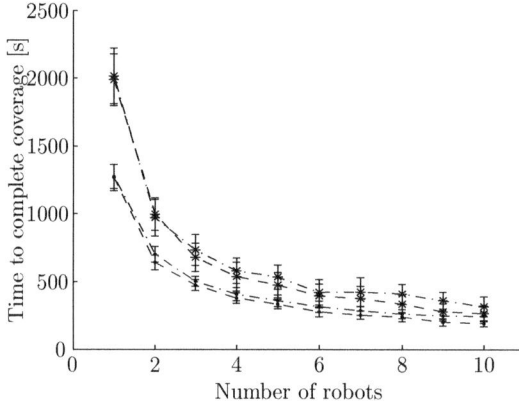

Figure 6.12: Prediction of the DES (– –) simulator and Webots (–··–) for wheel-slip of 10% (·) and 50% (∗) using a market-based coordination approach with re-auctioning provide close match and performance gracefully decays with increasing amount of slip noise.

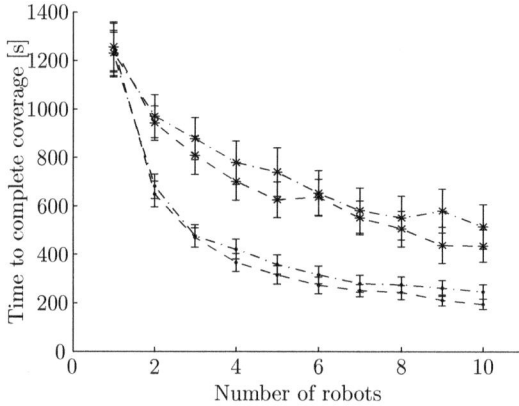

Figure 6.13: Prediction of the DES (– –) simulator and Webots (–··–) for global communication (·) and no communication (∗) using a market-based coordination approach with re-auctioning.

re-plan based on the location at which they arrived.

The effect os sensor and actuator noise could be well captured using the distribution of the time to inspect a blade and the probability to fail during navigation in the DES simulator when compared with Webots (Figures 6.8, 6.12, and 6.13) and real robot experiments (Figure 6.9).

The probability to fail is the same for all edges and all robots and the reliability of a robot is summarized by the parameters π^e and τ^v. In reality, some edges are more difficult to navigate than others, and navigation skills of robots might differ as well. In the turbine inspection case study this is imposed by the geometry of the blades, which require different behaviors for traversing an edge. In other scenarios and general cellular decompositions, the probability of successful edge traversal might be given by the terrain, robot capabilities, lighting conditions, or geometric constraints of the environment to name a few. Then, optimal coverage paths are not necessarily the shortest ones but those with the highest probability to lead to complete coverage.

6.6.2   Communication

Collaborative algorithms show improvement in time to completion even when the communication range is limited (Figure 6.10). As robots keep on meeting, information propagates slowly to all robots. This is an important finding (see also Jäger & Nebel 2002), as in most environments global communication cannot be guaranteed, either due to limitations on power consumption or due to environmental constraints that lead to communication loss; incidentally this is also the case for the miniature robotic platform and the case study being the motivation for the algorithms developed in this dissertation. Using communication is still beneficial even with a large localization error, i.e. exchanged information is wrong, albeit its benefit gets relatively smaller (Figure 6.11).

For large environments sharing the whole map obviously does not scale and robots would need to communicate only a subset of information every time progress is made. Here finding the right mix between communicating new and redundant information might be another interesting research problem.

6.7   Conclusion

This chapter presented a suite of increasingly complex deliberative algorithms for approximating an optimal partitioning of the environment, i.e. minimize time to completion. First, a non-collaborative, greedy algorithm is developed in which each robot follows provably complete trajectories by always moving towards the closest uncovered vertex on a local map. This algorithm does not require communication, and localization is achieved by node counting. The performance of this algorithm gracefully decays to the performance of a reactive version upon sensor and actuator noise (e.g., wheel-slip).

The performance can be drastically improved with communication and localization. In an implicit collaboration scheme, robots broadcast their coverage progress to other robots while using a global coordinate system as frame of reference. The experimental results from a realistic simulation and a discrete-event simulation show that the algorithms are robust with respect to unreliable communication and positional noise, and they gracefully decay to the performances of the non-collaborative and reactive versions of the algorithm, respectively.

Using explicit collaboration, a near-optimal partitioning of the environment is calculated using a market-based algorithm. For increasing robustness, the partitioning is continuously re-auctioned upon coverage progress. Upon loss of communication, this algorithm gracefully decays to the performance of a near-optimal single robot algorithm, and to that of a reactive approach on positional error.

The performance of all three algorithms is captured by probabilistic models that are carefully calibrated using the sensor and actuator noise characteristics measured from the real robotic platform. Due to the small state space, the performance of the non-collaborative algorithm without localization could be captured well by an analytical expression, whereas algorithms using communication and localization are modeled by sampling using discrete event simulation a reasonable amount (100 experiments per configuration in this chapter) of possible trajectories in the state space .

Chapter Summary

- Coverage performance can be significantly improved by sharing task progress among the robots, which allows the robots to minimize redundant coverage. This approach requires localization and radio communication.

- If the environment and robot positions are known beforehand, an optimal partitioning of the environment can be calculated. This task is $\mathcal{NP}$-hard and involves a solution of the Shortest-Path-Problem for all possible permutations. Near-optimal approximation can be achieved by a market-based algorithm, and continuous re-planning increases robustness against sensor and actuator noise.

- Upon communication failure or with communication subject to limited range, the coverage performance gradually decreases towards a non-collaborative version of the algorithm and remains provably complete if at least one robot does not fail. Upon an error of the localization mechanism, all algorithms gradually decay to a randomized algorithm. In this case, the environment needs to be covered redundantly (patrolling), and upper bounds for the minimal number of tours can be calculated based on the expected error in localization.

■ The performance of the presented algorithms is captured well by sampling possible trajectories in the state space of the system by using probabilities and delays measured on a real robot or realistic simulation. These kind of models can then be used for exploring different design parameters of the system.

CHAPTER 7

Comparing Coordination Schemes for Distributed Boundary Coverage

Designing coordination algorithms for distributed boundary coverage can be understood as a constraint optimization problem subject to size constraints, energy limitations, available sensors, and sensor and actuator noise present in the system. The system performance is then a trade-off between algorithmic requirements, such as completeness or a specific time to completion, and the available resources. Although this applies well to any engineering problem, it is most pertinent in miniature multi-robot systems, where all of the above constraints might be exceeded to their limits.

The preceding chapters focused on a series of reactive (Chapter 5) and deliberative (Chapter 6) algorithms. This chapter provides a quantitative (Section 7.1) and qualitative comparison of the different algorithms and analysis tools used. The algorithms are classified into hardware requirements and according to the benefits they provide to the user (Section 7.2). Although partly deliberative, none of the control approaches considered in this thesis are deterministic. Therefore, the performance analysis is achieved using probabilistic models, and Section 7.3 outlines similarities and differences in the analysis tools. Furthermore, it extracts the general properties of the models used that might well be useful for modeling other multi-robot systems.

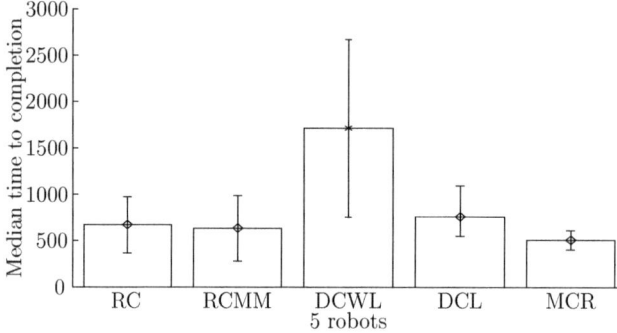

Figure 7.1: Median performance for 5 robots using reactive coverage without collaboration (RC), reactive coverage using mobile markers (RCMM), deliberative non-collaborative coverage without localization (DCWL), deliberative collaborative coverage with localization (DCL), and market-based coverage with re-auctioning (MCR) for 32, 34, 100, 9, 9 replications, respectively. Real robot experiments ($\Diamond$) and realistic simulation ($*$).

7.1   Quantitative Performance Comparison

Experimental results for 5 and 10 robots are depicted in Figures 7.1 and 7.2. When the experimental data for real robots is not available, results from Webots simulation are shown instead.

For the teams of 5 and 10 robots, reactive algorithms with and without mobile marker-based collaboration and the market-based algorithm achieve good absolute time-to-completion. The market-based algorithm also shows the absolute best performance for the 5 robot teams. The results of the reactive algorithms with and without collaboration are comparable, as the effects of local communication become only visible for higher densities of robots (compare also Figure 5.15). Deliberative coverage without localization showed the worst performance, due to the time-consuming navigation behaviors required by the deliberative algorithm, together with poor coordination due to the lack of localization and communication.

7.1.1   Scalability

This section discusses the scalability of the algorithms with respect to adding robots, removing robots, and to the environment size. All of the described algorithms allow for the addition and removal of robots at run-time. The performances of all algorithms

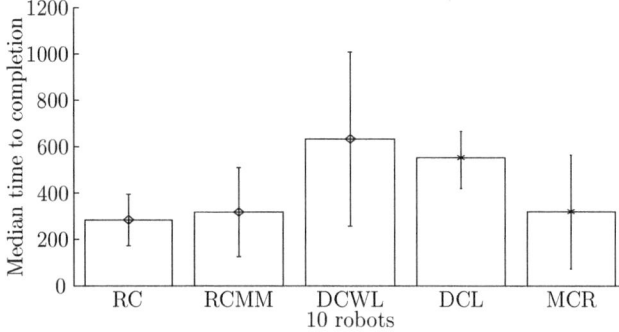

Figure 7.2: Median performance for 10 robots using reactive coverage without collaboration (RC), reactive coverage using mobile markers (RCMM), Deliberative Non-Collaborative Coverage without Localization (DCWL), Deliberative Collaborative Coverage with Localization (DCL), and Market based Coverage with Re-Auctioning (MCR) for 32, 34, 100, 9, 9 replications, respectively. Real robot experiments ($\Diamond$) and realistic simulation ($*$).

benefit from additional team-members, whereas reactive approaches benefit more from a large number of robots. The performances of all algorithms suffer from the removal of team-members, whereas the market-based approaches face an increased computational load as local bid computation (involving solving the TSP) becomes exponentially harder when the number of vertices per robot increases.

The effect of increasing the size of the environment for a fixed team size is investigated using microscopic models. Figure 7.3, *left*, shows median time-to-completion for 10 robots in the environments of 5x5, 16x16, and 50x50 cells (25, 256, and 2500 total area, respectively) and the 95% confidence interval (50 experiments each) for non-collaborative reactive coverage and mobile marker-based coverage obtained by synchronous, agent-based simulation. Figure 7.3, *right*, shows median time-to-completion for 10 robots in the same environments, but using deliberative collaborative coverage with localization.

The results show a tendency of a super-linear performance for both the reactive and the deliberative algorithm (time to completion is increased by around a factor of 5 whereas the area is increased by a factor of 10). This effect might have its roots in reduced redundant coverage for lower densities of robots. One can also see that the purely local communication in the reactive algorithms looses its effect in larger environments as the probability that two robots meet decreases.

The effect of increasing the number of robots has been investigated for deliberative

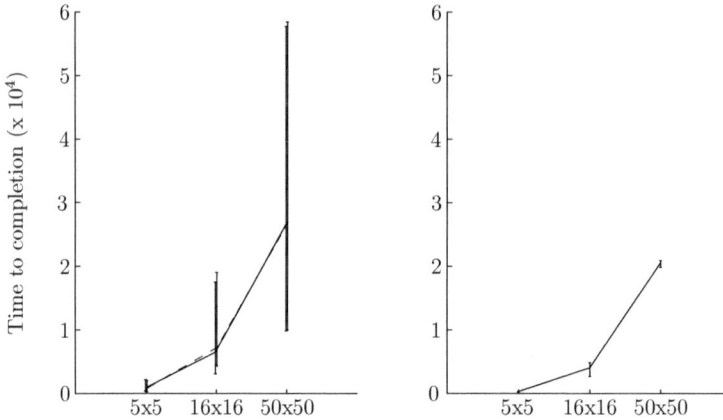

Figure 7.3: Median performance for 10 robots in environments spanning two orders of magnitude (25, 256, and 2500 cells). *Left*: Results for reactive mobile marker-based coverage (—) and non-collaborative reactive coverage (– –). *Right*: Results for deliberative collaborative coverage with localization (10% wheel-slip). Error bars depict the 95% confidence interval. 50 experiments using synchronous, agent-based simulation per configuration.

coverage with localization. Figure 7.4 shows the median time-to-completion for teams of 1 to 10 robots in the environments of 10x10, 16x16 and 25x25 cells. Increasing the number of robots increases the performance, whether the robots collaborate or not. This effect is maintained over two orders of magnitude of the environment. Figure 7.4 also shows that adding just a few robots already drastically improves performance.

All the algorithm's computational requirements scale well with the size of the environment, except for the market-based approaches. For large environments and small teams, the computation of bids in market-based approaches might become unfeasible due to the algorithmic complexity of bid computation. A remedy for this problem is to use a more coarse partitioning of the environment at cost of optimality.

7.1.2 Robustness

The amount of noise present in the Alice platform has been sufficient for creating essentially random patterns on collective level (except sweeping along a blade's contour with 50% chance) using the reactive algorithms presented in this dissertation. Thus, no recovery policies, except closed-loop reactive behaviors, are required. All the delibera-

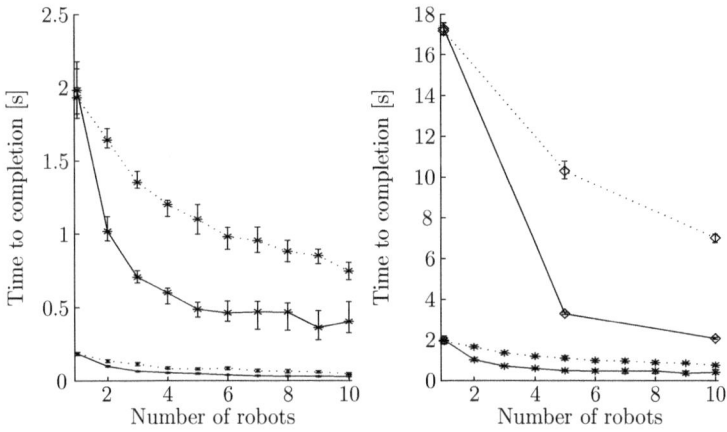

Figure 7.4: Median performance for 1–10 robots in environments spanning two orders of magnitude (25, 256 and 2500 cells). Results compare performance of deliberative collaborative coverage with localization for global communication (—) and no communication (· · ·) for environments of 5x5 (–.–), 16x16 (–∗–) (*left*), and 16x16 (–∗–) and 50x50 (–◇–) (*right*). 50 experiments using DES simulation per configuration.

tive approaches instead require some sort of recovery mechanism for dealing with robot failure. The deliberative algorithm without localization starts over from scratch after failure. The deliberative algorithm for coverage of unknown environments with localization plans on-line and thus automatically takes into account any updates to the robot's and the environmental state. Finally, the market-based algorithm requires constant re-auctioning of the partitioning for maintaining robustness. This tendency indicates a trade-off between algorithmic complexity and optimal performance as a function of sensor and actuator noise: additional computational effort for achieving "better" co-ordination might be counter-effected by necessary re-planning due to robot failure and uncertainty of the environment.

Results from Chapter 6 show that for unreliable or no communication the deliberative collaborative algorithms gracefully decay (see Figure 6.10 and Figure 6.13, respectively). Under absence of communication all deliberative algorithms will construct a minimal spanning tree of the environment by design, and will thus show comparable[1] behavior and performance as the non-collaborative deliberative algorithm presented in Section 6.2. Under the influence of sensor and actuator noise, all deliberative algorithms then gracefully decay to that of a randomized algorithm.

7.2 Comparing Performance as a Function of Requirements and Benefits

Algorithms can be compared as a function of the requirements on the individual robotic platform and the system, as well as a function of the benefits provided to the user. Requirements and benefits specific to the inspection case study are summarized in Tables 7.1 and 7.2, respectively. Requirements on the individual robotic platform usually also pose requirements on the system as a whole. So might it be unfeasible to mark the interior of the turbine or to provide some other signal for localization, or communication with the outside is impossible due to shielding. Some benefits are a by-product of the requirements posed by the algorithm. For instance, localization allows for mapping collected sensory information to a specific location. It is also for such benefits of a particular solution that a large number of simple robots cannot substitute a smaller team of more complex robots only because it provides the same performance to a particular metric.

---

[1]Whereas the algorithms from Sections 6.2 and 6.3 use a greedy policy comparable to a depth-first-search on the graph, the market-based algorithm calculates an optimal tour over the graph by solving the TSP.

		Local Communica-tion (infra-red)	Long-range communication (radio)	Localization with camera	External compu-tation	Known map
Reactive	Non-collaborative coverage	-	-	-	-	-
	Stationary marker-based coverage	✓	-	-	-	-
	Mobile marker-based cover-age	✓	-	-	-	-
Deliberative	Non-collaborative coverage without localization	-	-	-	-	-
	Collaborative coverage with localization	-	✓	✓	-	-
	Market-based coverage with re-auctioning	-	✓	✓	✓	✓

Table 7.1: Requirements for the reactive and deliberative algorithms for distributed coverage considered in this thesis. Requirements range from very limited sensors and actuators to long-range communication, availability of a map of the environment and global localization

		Progress Monitoring	Mapping of flaws	Low Power Consumption
Reactive	Non-collaborative coverage	-	-	+
	Stationary marker-based coverage	-	-	+
	Mobile marker-based coverage	-	-	+
Deliberative	Non-collaborative coverage without Localization	-	-	+
	Collaborative coverage with Localization	+	+	-
	Market-based coverage with re-auctioning	+	+	-

Table 7.2: Benefits of the reactive and deliberative algorithms. Some benefits come for "free" with the chosen coordination algorithm, and are difficult to achieve without the requirements of this algorithm. For instance, progress monitoring and mapping of flaws are resulting from localization and energy efficiency results from individual simplicity required by the random algorithms.

7.3   Modeling Robotic Swarms

The algorithms discussed in Chapter 5 are essentially memory-less. This property significantly reduces the state space of the robotic system: the current robot state is given at any time by the actual reactive control scheme uses (e.g., random walk, wall-following) and by the value of timers (e.g., time-out for acting as a stationary marker). Section 4.2.1 shows how a probabilistic model for such a system can be formulated and how delay states can be reduced to a single state. Similarly, the state of the environment (whether an element is covered or not), can be summarized by a single difference equation, assuming that the identity of the covered elements does not matter.

Depending on the assumptions on the algorithm, it is possible to reduce also controllers with memory to a tractable system. For instance, consider the following scenario. Robots follow the individual, reactive behavior described in Section 5.1.1, but store the id of every covered element (acquired by the camera). Upon encountering an element with a known id, a robot will immediately abandon coverage. Assuming that the abandonment behavior takes much less time than coverage, this policy might save considerable inspection time. Although the state space is large if the coverage state of every element in the environment would need to be maintained, the likelihood of encountering a covered element is simply given by the ratio of covered vs. un-covered elements, and can thus be captured by a single difference equation. Notice, that this would not be the case if the robot would not be random walking but deliberative move towards uncovered areas.

Indeed, a deliberative policy for distributed coverage, e.g. that of Section 6.2, requires a robot to keep track of its (relative) position with respect to its initial position and of the coverage progress. Due to the large number of possible robot and environmental states, a probabilistic model capturing the population dynamics is unsuitable. Instead, the environment can be modeled as a graph, which allows to formally describe the robot controller (e.g., as solution to a shortest path problem) and to gain analytical insight (e.g., length of a minimal Hamiltonian cycle). However, if the underlying reactive control schemes that execute trajectories calculated according to a deliberate control policy are unreliable (as it is in particular the case for miniature robots, and in some extent also for large robotic platforms), even deliberative robot controllers exhibit essentially probabilistic performance. This insight has already been formulated by Gat (1995). Gat argues that distance is not as correlated to time as usually assumed and that performance of a robot system is exponential distributed as opposed to following a normal distribution. These observations are explained by the non-determinism of reactive behavior that makes the time a robot travels for a particular distance unpredictable, and might lead to execution times that are significantly longer than the average, which explains the long-tailed distribution in robot performance.

In order for modeling this phenomenon, we propose (Correll & Martinoli 2007b) to *measure* the reliability of individual robot behaviors and the time they require for execution. By sampling from the distribution of these measurements, the robot controller, e.g. defined by (6.9), can then be simulated using a discrete event simulator (Rutishauser et al. 2007).

When recovery from a failed reactive behavior is impossible, the system can be modeled as a series of successful deliberative sequences that each contribute to a particular metric. In the coverage case study this is the case when either localization is not available (Section 6.2) or when localization is noisy (Section 6.3.3). For instance, considering the three algorithms from Sections 5.1, 6.2, and 6.3 (reactive coverage without collaboration, non-collaborative deliberative coverage without localization, and collaborative deliberative coverage with localization, respectively):

The algorithm described in Section 5.1 eventually achieves complete coverage by repeating a simple deliberative behavior, which is circumnavigating a blade until a time-out expires. The associated model, which achieves close agreement with experimental data is given by (5.7):

$$M_v(k+1) = M_v(k)\left(1 - p_e N_s(k)\right) \tag{7.1}$$

where $p_e N_s(k)$ are the average number of elements being covered during time interval k.

In contrast, the algorithm described in Section 6.2 achieves coverage by repeatedly starting over after a deliberative policy failed due to sensor and actuator noise. We then showed experimentally that coverage progress is well described by (6.15)

$$M_v(\kappa+1) = M_v(\kappa)\left(1 - \frac{\mu}{\|\mathcal{V}\|}\right)^{N_0} \tag{7.2}$$

where $\frac{\mu}{\|\mathcal{V}\|}$ is the average fraction of the environment a robot covers before failure and N_0 are the number of robots. The length of an iteration κ is given by the average time before failure ($\kappa = \mu \tau_v$).

Finally, when considering imperfect localization (Section 6.3), complete coverage cannot be guaranteed, and the fraction of the environment being covered is given by the lower bound $1 - p_f$, where p_f corresponds to the probability of localization failure. Coverage progress is then well modeled by the lower bound

$$M_v(K+1) = M_v(K)(1 - p_f) \tag{7.3}$$

The length of an iteration K is given by the time needed for the team to cover the environment at least once, i.e. perform one tour.

Thus, in (7.1)–(7.3) coverage progress is described by the same asymptotic dynamics although progressing at different speeds.

7.4 Conclusion

Designing an inspection system based on a robotic swarm consists of solving various trade-offs. The most pertinent being the option to replace a small team of highly capable robots by a larger team of simpler robots that provide the same or a similar performance. However, while this can be necessary as the simpler robots might offer longer autonomy due to less energy consumption, more capable robots often provide benefits that are difficult to obtain otherwise, e.g., localization, which is difficult to reconstruct off-line. For the boundary coverage case study, reactive solutions seem to offer a better performance with respect to time to completion, but there are no means for mapping of flaws and monitoring progress. Adding localization and communication to the robots yielded these benefits and enabled more deliberative algorithms. Comparing two different algorithms with drastically different theoretical performances showed that the performance is only marginally increased with near-optimal planning although this requires extensive computation and an a priori knowledge of the environment. This allows for deriving the following rules of thumb: more reasoning leads to a better performance. However, depending on the level of sensor and actuator noise, the improvement might be marginal and might not be worth the additionally required effort (additional hardware, computation, or communication). The probabilistic models allow for exploring these different scenarios and help in the design process by answering questions such as "What is the upper bound on communication loss that the system has to provide so that algorithm X provides an advantage over algorithm Y?".

Despite the use of deterministic control schemes, the performance of a real-robot is essentially probabilistic and completion asymptotic. This is due to the fact that reactive behaviors are based on potentially noisy sensor information and thus have unpredictable completion times. In systems where a deliberative policy can recover due to some sort of global reference, an accurate prediction of the system performance can be achieved by sampling from the distribution of execution time and the completion probability for each reactive behavior used by the higher-level deterministic algorithm. In systems where a deliberative policy cannot recover once a reactive behavior fails (i.e. task progress achieved so far cannot be reconstructed). An accurate prediction of the system performance can be achieved by modeling the system as a sequence of independent trials and by sampling from the distribution of the task contribution that a robot is expected to provide before failure. This concept is illustrated by three algorithms within increasing complex deliberative algorithms. The uncertainty due to a reactive policy (7.1), wheel-slip (7.2), or erroneous localization (7.3), leads in each case to a equation of similar form, which shows the same asymptotic behavior with a different expected time for convergence. We conjecture that any kind of uncertainty will reduce any coverage policy, collaborative or non-collaborative, reactive or deterministic, to probabilistic completeness.

Chapter Summary

- Algorithms for a multi-robot inspection task need to be compared not only based on time-to-completion, but also on the requirements on the individual robotic platform and on the other benefits for the user that a certain platform provides.

- Deliberative approaches are preferable over reactive approaches as they generally provide a higher degree of repeatability. The required accuracy in the navigation requirements might lead to an effectively faster completion of a reactive approach, however.

- The various design choices appearing in a multi-robot system can be addressed by modeling the system at multiple levels of abstraction, but they require detailed information on the system's sensor and actuator noise characteristics.

CHAPTER 8

Conclusion

This dissertation compares a series of coordination algorithms for a distributed boundary coverage case study. The algorithms range from reactive to deliberative solutions and they incrementally raise the requirements on the robotic platform. The coverage performance drastically benefits from reasoning, collaboration, and available a priori information. These benefits become smaller, however, with the increasing amount of sensor and actuator noise, which is inevitable even on larger robotic platforms. Thus, all algorithms are designed such that their performance will gracefully decay to that of a reactive algorithm with increasing sensor and actuator noise or the failure of some robots. Choosing a particular family of algorithms is then not only a function of the available robot capabilities and a priori information, but also a function of the reliability of the robotic platform. Although providing by far the best theoretical performance in terms of time to completion, a market-based approach comes with considerable cost to communication and computation. Given the strong sensor and actuator noise of the Alice platform, similar performance is achieved with a reactive solution, which requires a slightly larger number of potentially less expensive robots, but does not allow for proprioceptive monitoring of task progress.

Both reactive and deliberative algorithms are modeled by probabilistic models, which carefully model sensor and actuator noise that are at the root of non-determinism in a (multi-)robotic system. In a multi-robot system, it is likely that the overall system behavior comes close to the "average" behavior predicted by a probabilistic model for a single agent. By providing concepts for modeling swarms of deliberative agents, this thesis extends upon previous work on the probabilistic modeling of multi-robot systems, which consist mainly on fully reactive systems. Deliberation and memory lead to an explosion of the state space of the system, which in turn lead to an untractable number of master equations that maintain a probability distribution over all possible system states. In this case, the system can be analyzed by *simulating* the distributed system and averaging over a reasonable number of sample trajectories through state space. The simulation is implemented by carefully modeling the (deterministic) controller and by introducing randomized transitions that correspond to carefully calibrated random elements of the individual robotic platform or the environment. This can be achieved for instance by a discrete event system simulator.

In order to field multi-robot inspection systems, however, a series of technological and social hurdles still need to be overcome. The trend in industry is to enhance the manual systems currently in use instead of substituting them. The level of autonomy might then increase in small steps upon the acceptance of semi-autonomous solutions. This would also require rethinking concepts such as safety and liveness from a probabilistic perspective, as robotic systems embedded in the real world are — unlike software agents — intrinsically probabilistic. From a technological perspective, more research is necessary in sensor fusion, human-swarm interaction, and the synthesis of individual robot controllers based on the input of a human or an expert system.

On the spectrum from reactive coordination to centralized near-optimal coordination, small-scale multi-robot systems fielded in the near future will benefit most likely from one or more centralized components that perform near-optimal planning, rather than being fully distributed. Indeed, even for miniature robotic platforms of a cubic inch as considered in this thesis, such approaches are feasible and yield the best performances when compared with other, less coordinated approaches. Whenever the computational capabilities of the individual platform and the team objective allow, centralized control can also be distributed among the robot team (as illustrated with the market-based algorithm considered in this thesis) and decrease the potential vulnerability of solutions that rely on centralized entities.

On the contrary, reactive approaches — which have been shown to become analytically tractable using probabilistic models — have the potential for integration on even smaller platforms on the nano-meter scale and they provide better scalability with respect to the number of agents as no central control is required. However, the size of the drive train and battery are major bottlenecks when down-scaling robotic systems, rather than the implementation of small-scale computation and communication devices.

Acar, E., Choset, H., Zhang, Y. & Schervish, M. (2003), 'Path planning for robotic demi-
ning: Robust sensor-based coverage of unstructured environments and probabilistic
methods', *Int. J. of Robotics Research* **22**(7–8), 441–466.

Agassounon, W., Martinoli, A. & Easton, K. (2004), 'Macroscopic modeling of aggregation experiments using embodied agents in teams of constant and time-varying
sizes', *Autonomous Robots* **17**(2–3), 163–191. Special Issue on Swarm Robotics.

Albers, S. & Henzinger, M. (2000), 'Exploring unknown environments', *SIAM Journal
on Computing* **29**(4), 1165–1188.

Amstutz, P. (2007), Robust distributed coverage with a team of networked miniature
robots using market-based algorithms, Master's thesis, SWIS-MP7, École Polytechnique Fédérale Lausanne.

Andersson, M. & Sandholm, T. (2000), Contract type sequencing for reallocative negoti-
ation, *in* 'Proc. of the The 20th Int. Conference on Distributed Computing Systems
(ICDCS 2000)', Washington, DC, USA, pp. 154–161.

Arkin, R. (2000), *Behavior-Based Robotics*, 2nd edn, The MIT press, Cambridge, MA,
USA.

Asmussen, S. (1987), *Applied Probability and Queues*, Wiley, New York.

Balch, T. (1998), Behavioral Diversity in Learning Robot Teams, PhD thesis, Georgia
Institute of Technology, Atlanta, Georgia, USA.

Balluchi, A., Benvenuti, L., Engell, S., Geyer, T., Johansson, K. H., Lamnabhi-
Lagarrigue, F., Lygeros, J., Morari, M., Papafotiou, G., Sangiovanni-Vincentelli,

A. L., Santucci, F. & Stursberg, O. (2005), 'Hybrid control of networked embedded systems', *European Journal of Control* **11**, 1–31 (478–508).

Berhault, M., Huang, H., Keskinocak, P., Koenig, S., Elmaghraby, W., Griffin, P. & Kleywegt, A. (2003), Robot exploration with combinatorial auctions, *in* 'IEEE/RSJ Int. Conf. on Intelligent Robots and Systems (IROS)'.

Berman, S., Halasz, A., Kumar, V. & Pratt, S. (2006), Algorithms for the analysis and synthesis of a bio-inspired swarm robotic system, *in* 'Proc. of the SAB 2006 Workshop on Swarm Robotics', Rome, Italy.

Bonabeau, E., Dorigo, M. & Theraulaz, G. (1999), *Swarm Intelligence: From Natural to Artificial Systems*, SFI Studies in the Science of Complexity, Oxford University Press, New York, NY, USA.

Braitenberg, V. (1986), *Vehicles: Experiments in Synthetic Psychology*, The MIT Press.

Bronée, S. (2005), Collaborative, GPS-free techniques for localization in miniature robots, Semester project SWIS-SP5, École Polytechnique Fédérale Lausanne.

Burgard, W., Moors, M., Stachniss, C. & Schneider, F. (2005), 'Coordinated multi-robot exploration', *IEEE Transactions on Robotics* **21**(3), 376–378.

Butler, Z., Rizzi, A. & Hollis, R. (2001), Complete distributed coverage of rectilinear environments, *in* 'Int. Workshop on Algorithmic Foundations of Robotics (WAFR)', Boston, MA, USA.

Camazine, S., Deneubourg, J.-L., Franks, N. R., Sneyd, J., Theraulaz, G. & Bonabeau, E. (2001), *Self-Organization in Biological Systems*, Princeton Studies in Complexity, Princeton University Press.

Caprari, G. & Siegwart, R. (2005), Mobile micro-robots ready to use: Alice, *in* 'IEEE/RSJ Int. Conf. on Intelligent Robots and Systems (IROS)', Edmonton, Alberta, Canada, pp. 3295–3300.

Cassandras, C. (1993), *Discrete Event Systems — Modeling and Performance Analysis*, Aksen Associates Incorporated Publishers.

Chaimowicz, L., Michael, N. & Kumar, V. (2005), Controlling swarms of robots using interpolated implicit functions, *in* 'IEEE Int. Conf. on Robotics and Automation (ICRA)', Barcelona, Spain, pp. 2487–2492.

Choset, H. (2000), 'Coverage of known spaces: The boustrophedon cellular decomposition', *Autonomous Robots* **9**, 247–253.

Choset, H. (2001), 'Coverage for robotics—a survey of recent results', *Annals of Mathematics and Artificial Intelligence* **31**, 113–126.

Cianci, C., Raemy, X., Pugh, J. & Martinoli, A. (2006), Communication in a swarm of miniature robots: The e-Puck as an educational tool for swarm robotics, *in* 'Proc. of the SAB 2006 Workshop on Swarm Robotics', Rome, Italy.

Correll, N., Cianci, C., Raemy, X. & Martinoli, A. (2006), Self-Organized Embedded Sensor/Actuator Networks for "smart" Turbines, *in* 'IROS 2006 Workshop: Network Robot System: Toward intelligent robotic systems integrated with environments', Beijing, China.

Correll, N. & Martinoli, A. (2004a), Collective inspection of regular structures using a swarm of miniature robots, *in* 'Proc. of the Int. Symp. on Experimental Robotics (ISER)', Springer Tracts in Advanced Robotics (STAR), Vol. 21, 2006, Singapore, pp. 375–385.

Correll, N. & Martinoli, A. (2004b), Modeling and optimization of a swarm-intelligent inspection system, *in* 'Proc. of the Int. Symp. on Distributed Autonomous Robotic Systems (DARS)', Springer Distributed Autonomous Robotic Systems (2007), Toulouse, France, pp. 369–378.

Correll, N. & Martinoli, A. (2005), Modeling and analysis of beacon-based and beaconless policies for a swarm-intelligent inspection system, *in* 'IEEE Int. Conf. on Robotics and Automation (ICRA)', Barcelona, Spain, pp. 2488–2493.

Correll, N. & Martinoli, A. (2006a), System identification of self-organized robotic swarms, *in* 'Proc. of the Int. Symp. on Distributed Autonomous Robotic Systems (DARS)', Springer Distributed Autonomous Robotic Systems, Minneapolis, MN, USA (2006), pp. 31–40.

Correll, N. & Martinoli, A. (2006b), Towards optimal control of self-organized robotic inspection systems, *in* '8th Int. IFAC Symp. on Robot Control (SYROCO)', Bologna, Italy.

Correll, N. & Martinoli, A. (2007a), Modeling self-organized aggregation in a swarm of miniature robots, *in* 'Int. Conf. on Robotics and Automation, Workshop on Collective Behaviors inspired by Biological and Biochemical Systems', Rome, Italy.

Correll, N. & Martinoli, A. (2007*b*), Robust distributed coverage using a swarm of minia-ture robots, *in* 'IEEE Int. Conf. on Robotics and Automation (ICRA)', Rome, Italy, pp. 379–384.

Correll, N., Rutishauser, S. & Martinoli, A. (2006), Comparing coordination schemes for miniature robotic swarms: A case study in boundary coverage of regular structures, *in* 'Proc. of the Int. Symp. on Experimental Robotics (ISER)', Springer Tracts in Advanced Robotics (STAR), to appear, Rio de Janeiro, Brazil.

Correll, N., Sempo, G., de Meneses, Y. L., Halloy, J., Deneubourg, J.-L. & Martinoli, A. (2006), SwisTrack: A tracking tool for multi-unit robotic and biological research, *in* 'IEEE/RSJ Int. Conf. on Intelligent Robots and Systems (IROS)', Beijing, China, pp. 2185–2191.

Cortés, J., Martínez, S., Karatas, T. & Bullo, F. (2004), 'Coverage control for mobile sensing networks', *IEEE Transactions on Automatic Control* **20**(2), 243–255.

Cruz, A. & Ribeiro, M. (2005), 'Robtank inspec - in service robotized inspection tool for hazardous products storage tanks', *Industrial Robot: An International Journal* **32**(2), 157–162.

Dias, M. B. & Stentz, A. (2000), A free market architecture for distributed control of a multi-robot system, *in* 'Proc. of the 6th Int. Conf. on Intelligent Autonomous Systems', Venice, Italy, pp. 115–122.

Dias, M., Zlot, R., Kalra, N. & Stentz, A. (2006), 'Market-based multirobot coordination: a survey and analysis', *Proc. of the IEEE* **94**(7), 1257–1270. Special Issue on Multi-Robot Systems.

Easton, K. & Burdick, J. (2005), A coverage algorithm for multi-robot boundary inspec-tion, *in* 'IEEE Int. Conf. on Robotics and Automation (ICRA)', Barcelona, Spain, pp. 727–734.

Federal Aviation Administration (1998), Acceptable methods, techniques, and practices — aircraft inspection and repair, Technical Report AC 43.13-1B, Federal Aviation Administration AFS-640.

Fischer, W., Tâche, F. & Siegwart, R. (2007), Magnetic wall climbing robot for thin surfaces with specific obstacles, *in* 'Proceedings of the International Symposium on Field and Service Robotics (FSR)', Chamonix, Switzerland. To Appear.

Friedrich, M., Galbraith, W. & Hayward, G. (2006), Autonomous mobile robots for ultrasonic NDE, *in* 'Proceedings of the IEEE Ultrasonics Symposium', pp. 902–905.

Gabriely, Y. & Rimon, E. (2001), 'Spanning-tree based coverage of continuous areas by a mobile robot', *Annals of Mathematics and Artificial Intelligence* **31**(1–4), 77–98.

Garg, S. (2004), Controls and health management technologies for intelligent aerospace propulsion systems, *in* '42nd Aerospace Sciences Meeting and Exhibition', Reno, NV, USA. AIAA paper: 2004-949.

Gat, E. (1995), 'Towards principled experimental study of autonomous mobile robots', *Autonomous Robots* **2**(3), 179–189.

Gazi, V. & Fidan, B. (2005), 'Swarm aggregations using artificial potentials and sliding mode control', *IEEE Transactions on Robotics* **21**(6), 1208–1214.

Gerkey, B. & Matarić, M. (2004), 'A formal analysis and taxonomy of task allocation in multi-robot systems', *Int. J. of Robotics Research* **23**(9), 939–954.

Gibson, M. & Bruck, J. (2000), 'Efficient exact stochastic simulation of chemical systems with many species and many channels', *Journal of Physical Chemistry A* **104**, 1876–1889.

Gillespie, D. (1977), 'Exact stochastic simulation of coupled chemical reactions', *The Journal of Physical Chemistry* **81**(25), 2340–2361.

Hazon, N. & Kaminka, G. (2005), Redundancy, efficiency, and robustness in multi-robot coverage, *in* 'IEEE Int. Conf. on Robotics and Automation (ICRA)', Barcelona, Spain, pp. 735–741.

Hazon, N., Mieli, F. & Kaminka, G. (2006), Towards robust on-line multi-robot coverage, *in* 'IEEE Int. Conf. on Robotics and Automation (ICRA)', Orlando, FL, USA, pp. 1710–1715.

Hinic, V., Petriu, E. & Whalen, T. (2007), Human-computer symbiotic cooperation in robot-sensor networks, *in* 'Proc. of the IEEE Instrumentation and Measurement Technology Conference', pp. 1–6.

Howard, A., Parker, L. & Sukhatme, G. (2006), 'The SDR experience: Experiments with a large-scale heterogeneous mobile robot team', *Int. J. of Robotics Research* .

Hsieh, M. & Kumar, V. (2006), Pattern generation with multiple robots, *in* 'IEEE Int. Conf. on Robotics and Automation (ICRA)', Orlando, FL, USA, pp. 2442–2447.

Hsieh, M., Loizou, S. & Kumar, V. (2007), Stabilization of multiple robots on stable orbits via local sensing, *in* 'IEEE Int. Conf. on Robotics and Automation (ICRA)', Rome, Italy, pp. 2312–2317.

Hunter, G. (2003), Morphing, self-repairing engines: A vision for the intelligent engine of the future, *in* 'AIAA/ICAS Int. Air & Space Symposium'.

Hybinette, M., Kraemer, E., Xiong, Y., Matthews, G. & Ahmed, J. (2006), SASSY: A design for a scalable agent-based simulation system using a distributed discrete event infrastructure, *in* 'Proc. of the 2006 Winter Simulation Conference', pp. 926–934.

Jadbabaie, A., Lin, J. & Morse, A. S. (2003), 'Coordination of groups of mobile autonomous agents using nearest neighbor rules', *IEEE Transactions on Automatic Control* **48**(6), 988–1001.

Jäger, M. & Nebel, B. (2002), Dynamic decentralized area partitioning for cooperating cleaning robots, *in* 'IEEE Int. Conf. on Robotics and Automation (ICRA)', Washington, DC, USA, pp. 3577–3582.

Johansson, R. (1993), *System Modeling and Identification*, Prentice Hall.

Kalra, N. & Martinoli, A. (2006), A comparative study of market-based and threshold-based task allocation, *in* 'Proc. of the Int. Symp. on Distributed Autonomous Robotic Systems (DARS)', Springer Distributed Autonomous Robotic Systems, Minneapolis, MN, USA (2006), pp. 31–40.

Krasowski, M., Greer, L. & Oberle, L. (2002), Mobile sensor technologies being developed, *in* 'Research and Technology', NASA Glenn Center, pp. 83–84.

Kumar, V., Rus, D. & Singh, S. (2004), 'Robot and sensor networks for first responders', *PERVASIVE Computing* pp. 24–33.

Lagoudakis, M., Markakis, V., Kempe, D., Keskinocak, P., Koenig, S., Kleywegt, A., Tovey, C., Meyerson, A. & Jain, S. (2005), Auction-based multi-robot routing, *in* 'Robotics: Science and Systems', Cambridge, MA, USA.

Lerman, K. & Galstyan, A. (2002), 'Mathematical model of foraging in a group of robots: Effect of interference', *Autonomous Robots* **2**(13), 127–141.

Lerman, K., Galstyan, A., Martinoli, A. & Ijspeert, A.-J. (2001), 'A macroscopic analytical model of collaboration in distributed robotic systems', *Artificial Life* **7**(4), 375–393.

Lerman, K., Jones, C., Galstyan, A. & Matarić, M. (2006), 'Analysis of dynamic task allocation in multi-robot systems', *Int. J. of Robotics Research* **25**(4), 225–242.

Lerman, K., Martinoli, A. & Galystan, A. (2005), A review of probabilistic macroscopic models for swarm robotic systems, *in* 'Proc. of the SAB 2004 Workshop on Swarm Robotics', Santa Monica, CA, USA, pp. 143–152. Lecture Notes in Computer Science Vol. 3342, Springer-Verlag, Berlin.

Li, L., Martinoli, A. & Abu-Mostafa, Y. (2004), 'Learning and Measuring Specialization in Collaborative Swarm Systems', *Adaptive Behavior* **12**(3–4), 199–212.

Litt, J., Wong, E., Krasowski, M. & Greer, L. (2003), Cooperative multi-agent mobile sensor platforms for jet engine inspection - concept and implementation, *in* 'IEEE Int. Conf. on Integration of Knowledge Intensive Multi-Agent Systems', pp. 716–721.

Liu, Y., Passino, K. & Polycarpou, M. (2003), 'Stability analysis of m-dimensional asynchronous swarms with a fixed communication topology', *IEEE Transactions on Automatic Control* **48**(1), 76–95.

Ljung, L. (1999), *System Identification—Theory for the User*, Prentice Hall.

Martinoli, A., Easton, K. & Agassounon, W. (2004), 'Modeling of swarm robotic systems: A case study in collaborative distributed manipulation', *Int. J. of Robotics Research* **23**(4), 415–436.

Martinoli, A., Ijspeert, A. J. & Mondada, F. (1999), 'Understanding collective aggregation mechanisms: From probabilistic modelling to experiments with real robots', *Robotics & Autonomous Systems* **29**, 51–63. Special Issue on Distributed Autonomous Robotic Systems.

McLurkin, J. & Smith, J. (2006), *Distributed Autonomous Robots 6*, Springer Verlag, Tokyo, chapter Distributed Algorithms for Dispersion in Indoor Environments using a Swarm of Autonmous Mobile Robots. To Appear.

Melcher, K. & Kypuros, J. (2003), Toward a fast-response active turbine tip clearance control, Technical Report NASA/TM2003-212627, NASA Glenn Research Center,

Cleveland, OH, USA. Proceedings of the 16th International Symposium on Airbreathing Engines.

Michel, O. (2004), 'Webots: Professional mobile robot simulation', *Journal of Advanced Robotic Systems* **1**(1), 39–42.

Milutinovic, D. & Lima, P. (2006), 'Modeling and centralized optimal control of a large-size robotic population', *IEEE Transactions on Robotics* **22**(6), 1280–1285.

Minar, N., Burkhart, R., Langton, C. & Askenazi, M. (1996), The Swarm Simulation System, a Toolkit for Building Multi-Agent Simulations, Working Paper 96-06-042, Santa Fe Institute, New Mexico, USA.

Olfati-Saber, R. & Murray, R. (2004), 'Consensus problems for networks of dynamic agents with switching topology and time-delays', *IEEE Transactions on Automatic Control* **49**, 1520–1533.

Payton, D., Estkowski, R. & Howard, M. (2003), 'Compound behaviors in pheromone robotics', *Robotics & Autonomous Systems* **44**, 229–240.

Petriu, E., Whalen, T., Abielmona, R. & Stewart, A. (2004), 'Robotic sensor agents', *IEEE Instrumentation & Measurement Magazine* **7**(3), 46–51.

Polastre, J., Szewczyk, R. & Culler, D. (2005), Telos: Enabling ultra-low power wireless research, *in* 'IEEE/ACM Int. Conf. on Information Processing in Sensor Networks (IPSN-SPOTS)', Los Angeles, CA, USA.

Prorok, A. (2006), Multi-level modeling of swarm robotic inspection, Master's thesis, SWIS-MP1, École Polytechnique Fédérale Lausanne.

Rathinam, M., Petzold, L., Cao, Y. & Gillespie, D. (2003), 'Stiffness in stochastic chemically reacting systems: The implicit tau-leaping method', *Journal of Chemical Physics* **119**(24), 12784–12794.

Reif, J. & Wang, H. (1999), 'Social potential fields: A distributed behavioral control for autonomous robots', *Robotics & Autonomous Systems* **27**, 171–194.

Rekleitis, I., Lee-Shue, V., New, A. P. & Choset, H. (2004), Limited communication, multi-robot team based coverage, *in* 'IEEE Int. Conf. on Robotics and Automation (ICRA)', Vol. 4, New Orleans, LA, USA, pp. 3462–3468.

Rekleitis, I., New, A. & Choset, H. (2005), Distributed coverage of unknown/unstructured environments by mobile sensor networks, *in* A. C. Schultz, L. E. Parker & F. Schneider, eds, '3rd International NRL Workshop on Multi-Robot Systems', Kluwer, Washington, D.C., pp. pages 145–155.

Ren, W., Beard, R. & Atkins, E. (2005), A survey of consensus problems in multi-agent coordination, *in* 'American Control Conference', Portland, OR, USA, pp. 1859–1865.

Riley, P. & Riley, G. (2003), SPADES — a distributed agent simulation environment with software-in-the-loop execution, *in* 'Proc. of the Winter Simulation Conference', pp. 817–825.

Rudnick, J. & Gaspari, G. (2004), *Elements of the Random Walk*, Cambridge University Press, Cambridge UK.

Rutishauser, S. (2007), Collaborative exploration and coverage with a team of networked miniature robots, Master's thesis, SWIS-MP8, École Polytechnique Fédérale Lausanne.

Rutishauser, S., Correll, N. & Martinoli, A. (2007), 'Collaborative coverage using a swarm of networked miniature robots', *Robotics & Autonomous Systems* . Submitted.

Sánchez, J., Vázquez, F. & Paz, E. (2005), Machine vision guidance system for a modular climbing robot used in shipbuilding, *in* 'Proc. of the Int. Conf. on Climbing and Walking Robots and the Support Technologies for Mobile Machines (CLAWAR 2005)', Springer Verlag Berlin, 2006, pp. 893–900.

Schwager, M., McLurkin, J. & Rus, D. (2006), Distributed coverage control with sensory feedback for networked robots, *in* 'Robotics: Science and Systems', Cambridge, USA.

Selten, R. (1975), 'A reexamination of the perfectness concept for equilibrium points in extensive games', *Int. Journal of Game Theory* **4**, 25–55.

Siegel, M. & Gunatilake, P. (1997), Remote inspection technologies for aircraft skin inspection, *in* 'IEEE Workshop on Emergent Technologies and Virtual Systems for Instrumentation and Measurement', Niagara Falls, Ontario, Canada, pp. 69–78.

Song, X., Wu, X. & Kang, Y. (2004), 'An inspection robot for boiler tube using magnetic flux leakage and ultrasonic methods', *Insight — Non-Destructive Testing and Condition Monitoring* **46**(5), 275–278.

Strogatz, S. (2000), *Nonlinear Dynamics and Chaos. With Applications to Physics, Biology, Chemistry, and Engineering*, Perseus Books Publishing.

Sugawara, K., Sano, M., Yoshihara, I. & Abe, K. (1998), 'Cooperative behavior of interacting robots', *Artificial Life and Robotics* **2**, 62–67.

Svennebring, J. & Koenig, S. (2004), 'Building terrain-covering ant robots', *Autonomous Robots* **16**(3), 313–332.

Tâche, F., Fischer, W., Siegwart, R., Moser, R. & Mondada, F. (2007), Adapted magnetic wheel unit for compact robots inspecting complex shaped pipe structures, *in* 'Proceedings of the 2007 IEEE/ASME International Conference on Advanced Intelligent Mechatronics', Zurich, Switzerland. To appear.

Tanner, H., Jadbabaie, A. & Pappas, G. (2005), *Flocking in teams of nonholonomic agents*, Vol. 309 of *Lecture Notes in Control and Information Sciences*, Springer Verlag, pp. 229–239.

Vaughan, R. & Gerkey, B. (2007), *Really Reusable Robot Code and the Player/Stage Project*, Vol. 30 of *Springer Tracts in Advanced Robotics*, Springer Verlag. To appear.

Williams, K. & Burdick, J. (2006), Multi-robot boundary coverage with plan revision, *in* 'IEEE Int. Conf. on Robotics and Automation (ICRA)', Orlando, FL, USA, pp. 1716–1723.

Wong, E. & Litt, J. (2004), Autonomous multi-agent robotics for inspection and repair of propulsion systems, *in* 'AIAA 1st Intelligent Systems Technical Conf.'.

Zheng, X., Jain, S., Koenig, S. & Kempe, D. (2005), Multi-robot forest coverage, *in* 'IEEE/RSJ Int. Conf. on Intelligent Robots and Systems (IROS)', Edmonton, Alberta, Canada, pp. 3852–3857.

Zlot, R. & Stentz, A. (2006), 'Market-based multirobot coordination for complex tasks', *Int. J. of Robotics Research* **25**(1), 73–102. Special Issue on the 5th International Conference on Field and Service Robotics.

Zlot, R., Stentz, A., Dias, M. & Thayer, S. (2002), Multi-robot exploration controlled by a market economy, *in* 'IEEE Int. Conf. on Robotics and Automation (ICRA)', Washington, DC, USA, pp. 3016–3023.

After earning his "Vordiplom" from the Technical University of Munich (Technische Universität München) in 2000, Nikolaus Correll graduated with a master's degree in Electrical Engineering from the Swiss Federal Institute of Technology Zurich (Eidgenössische Technische Hochschule Zürich, ETHZ) in Spring 2003. During his master's studies, he spent a term at the Lund Institute of Technology (Lunds Tekniska Högskola), Sweden, as an exchange student at the Department of Automatic Control. Nikolaus wrote his master's thesis at the Collective Robotics Group at the California Institute of Technology, Pasadena, CA, USA, about collaborative coverage.

After receiving his Master's degree, Nikolaus worked as a research assistant in the Collective Robotics Group at Caltech, which relocated shortly after to the Swiss Federal Institute of Technology Lausanne (École Polytechnique Fédérale Lausanne, EPFL) under its new name "Swarm-Intelligent Systems Group", where Nikolaus pursued graduate studies in Communication Systems. In November 2007, Nikolaus joined the Distributed Robotics Laboratory at the Massachusetts Institute of Technology, Cambridge, MA, USA, as post-doctoral fellow.

Nikolaus attended classes in Machine Learning, Swarm Intelligence, Optimal Control, and Dynamical Graphs at EPFL. In summer 2005, Nikolaus participated at the 2nd EURON/GEOPLEX Summer School on Modeling and Control of Complex Dynamical Systems at the University of Bologna, Italy.

Nikolaus work was awarded with a "Best Paper Award" at the 8th International Symposium on Distributed Autonomous Robotic Systems (DARS), by a "Student Travel Fellowship Award" at the International Federation of Robotics Research (IFRR) 10th International Symposium on Experimental Robotics, and received international media coverage in television (ARTE France/Germany, among others) and print (National Geographic, Scientific Computing, among others).